Global Data Shock

Strategic Ambiguity, Deception, and Surprise
in an Age of Information Overload

Robert Mandel

Stanford University Press
Stanford, California

Stanford University Press
Stanford, California

Printed in the United States of America on acid-free, archival-quality paper

Library of Congress Cataloging-in-Publication Data

Names: Mandel, Robert, author.
Title: Global data shock : strategic ambiguity, deception, and surprise in an age of
    information overload / Robert Mandel.
Description: Stanford, California : Stanford University Press, 2019. | Includes
    bibliographical references and index.
Identifiers: LCCN 2018052654 (print) | LCCN 2018056416 (ebook) | ISBN 9781503608979
    (e-book) | ISBN 9781503608252 (cloth; alk. paper) | ISBN 9781503608962 (pbk.; alk. paper)
Subjects: LCSH: National security. | Security, International. | Intelligence service. |
    Information resources management. | Disinformation.
Classification: LCC UA10.5 (ebook) | LCC UA10.5 .M327 2019 (print) | DDC 355/.033—dc23
LC record available at https://lccn.loc.gov/2018052654

Typeset by Newgen in 10/14 Minion

Cover design: Christian Fuenfhausen

Information Overload

A wealth of information creates a poverty of attention.
—Herbert A. Simon (American decision theorist), "Designing Organizations for an Information-Rich World"

It's not information overload. It's filter failure.
—Clay Shirky (American Internet expert), "It's Not Information Overload; It's Filter Failure"

There's a danger that too much stuff cramming in on people's minds is just as bad for them as too little, in terms of the ability to understand, to comprehend.
—Bill Clinton (American president), quoted in Todd S. Purdum, "Clinton Plans to Lift Public out of 'Funk'"

Ambiguity

There is no greater impediment to the advancement of knowledge than the ambiguity of words.
—Thomas Reid (Scottish philosopher), *Inquiry and Essays*

Accepting that the world is full of uncertainty and ambiguity does not and should not stop people from being pretty sure about a lot of things.
—Julian Baggini (British philosopher), "What Is This Foolish Lust for Uncertainty?"

Speech was given to man to disguise his thoughts.
—Charles Maurice de Talleyrand (French statesman and diplomat), quoted in Bertrand Berère, *Memoirs of Bertrand Berère*, vol. 4

Deception

In time of war, when truth is so precious, it must be attended by a bodyguard of lies.
—**Winston Churchill (British prime minister),**
***The Second World War*, vol. 5**

Though fraud in other activities be detestable, in the management of war it is laudable and glorious, and he who overcomes an enemy by fraud is as much to be praised as he who does so by force.
—**Niccolò Machiavelli (Italian historian and statesman),**
The Art of War

Life is the art of being well deceived; and in order that the deception may succeed it must be habitual and uninterrupted.
—**William Hazlitt (British writer), *The Round Table***

Surprise

War is, at first, the hope that one will be better off, then, the expectation that the other fellow will be worse off, then, the satisfaction that he isn't any better off, and finally, the surprise at everyone's being worse off.
—**Karl Kraus (Austrian writer), *Half-Truths and***
One-and-a-Half Truths

Always mystify, mislead, and surprise the enemy, if possible.
—**Stonewall Jackson (American soldier), quoted in John D.**
Imboden, "Stonewall Jackson in the Shenandoah"

Man [has] a limited capacity for change. When this capacity is overwhelmed, the consequence is future shock.
—**Alvin Toffler (American futurist), *Future Shock***

Contents

Figures and Tables

Figures

Tables

Acknowledgments

This study—my fifteenth book—has been incredibly enjoyable to ponder and write because it focuses on the complexities many countries' national governments and private citizens face in coping with a dramatically different quantity and quality of information compared to the past. Many of us feel overwhelmed by the scope and pace of change in data we perceive as relevant to our lives—unable to rapidly and correctly interpret trends or to distinguish what's true from what's false.

I dedicate this book to those in (1) the intelligence community tasked with interpreting foreign information, sorting through ambiguous, deceptive, and surprising data, and (2) the data science community driven to find new and better ways to derive useful insights from incoming information. I am deeply indebted to my undergraduate student research assistant Micael Lonergan for all her truly excellent work on this project. I also appreciate the ideas received from computer scientists and from government defense and intelligence officials. However, I alone take full responsibility for any disarming distortions or egregious errors found here.

Global Data Shock

Introduction

THIS BOOK EXPLORES two intertwined central global puzzles:

- How does information overload generally affect strategic ambiguity, deception, and surprise, and when does information overload seem most and least likely to increase the chances of strategic ambiguity, deception, and surprise?

- How do strategic ambiguity, deception, and surprise generally affect global security, and when do these manipulation techniques seem to have their most positive and negative impacts on global security?

Many public officials and private citizens sweepingly conclude that, regarding data acquisition, more is always better for decision legitimacy and effectiveness, and, regarding data transmission, complete clarity, transparency, and predictability are always better for global security. In contrast, this investigation somewhat counterintuitively finds that today's information overload frequently facilitates strategic ambiguity, deception, and surprise, challenging enlightened internationalist expectations of emerging global trust and cooperation. Increasing information access can dramatically worsen the signal-to-noise ratio,[1] eroding effective management of global intelligence and security challenges; operating as if most global communication is honest and accurate can have disastrous foreign policy consequences. This disconnect between common information and communication premises and existing realities results in *global data shock*, impeding both public officials' and private citizens' ability to interpret, respond to, and ultimately shape the world around them. In exploring circumstances affecting information overload's impact on strategic manipulation and on global security, and in recommending policies to manage global data shock, this analysis serves as a corrective to rosy assessments surrounding traditional "big data" analysis solutions and as a reminder that even with more information and better fact-checking and data assessment tools, today we may be out of touch with the world around us.

Analytical Scope

Given the breadth and depth of information and communication distortions, this book's scope is carefully circumscribed. This study considers the contrasting perspectives of both initiators and targets of strategic ambiguity, deception, and surprise because information misinterpretation and manipulation are embedded in two-way communication, for which responsibility is shared, making the quest to find fault or isolate who is right seems fruitless. Ten relevant case studies are explored: the 2017 foreign security policy style of American president Donald Trump; the 2016 "Brexit" vote to leave the European Union; the 2002–2003 nondiscovery of weapons of mass destruction in Iraq; the 2014 Russian annexation of Crimea; the 2011 Fukushima nuclear disaster; the 2008 Russian invasion of Georgia; the 2007 Israeli attack on the Syrian al-Kibar nuclear plant; the 2005 Andijan massacre in Uzbekistan; the 2001 al-Qaeda terrorist attacks on the United States; and the 1990 Iraqi attack on Kuwait.

This investigation's geographical scope is explicitly global, because the interplay between information overload and strategic manipulation cuts across national boundaries. This study emphasizes the post–Cold War time period, particularly the twenty-first century, for that is when the Internet-age information revolution transformed foreign data interpretation and manipulation. This work focuses on cross-state security issues (even when manipulation occurs within countries) because the most critical interpretation and manipulation costs and benefits are both international and security-oriented. Finally, this analysis concentrates on intentional and premeditated strategic manipulation applications because this deliberate planned use seems to have greater potential for improved future management. Overall, this integrated exploration aims to understand fully the complex web of causes, consequences, and cures surrounding global data shock.

Provocative and Unique Qualities

This book is both controversial and distinctive in several ways:

- It challenges the reliability, validity, and credibility of much open-source quantitative foreign security data.
- It challenges the value of always acquiring more information as a means of correctly interpreting situations and coming up with appropriate policies.
- It challenges the utility of traditional "big data" analysis in managing foreign security information.
- It challenges exclusive reliance on internal experts—with their biases and grooved thinking—as a means of correctly interpreting incoming information.
- It challenges the universal desirability of clarity, transparency, and predictability in global security communication.
- It is the first book to link information overload to changes in strategic ambiguity, deception, and surprise, considering the perspectives of both initiators and victims.
- It is the first book to detail a comprehensive set of global Information Age case studies presenting new material about the role of information overload and strategic manipulation.
- It is the first book to comprehensively explore the circumstances under which information overload most promotes foreign strategic manipulation and global insecurity.
- It is the first book to comprehensively present policy recommendations for constraining the negative security consequences of information overload and strategic manipulation.
- It is the first book to stress how both information overload and strategic ambiguity, deception, and surprise are critical concerns for citizens and government officials alike.

This study is unique not only in undertaking an integrated and timely analysis of global data shock but also in raising critical broader security concerns, including (1) global value divides, where despite growing globalization, tensions surround cultural diversity (including nativism, xenophobia, Islamophobia, and Western–non-Western and Global North–Global South frictions); (2) foreign intelligence failures, where despite massive data collection, many security aims are not reliably attained; (3) mass public frustration, where despite expanded news access, many citizens misunderstand foreign security policy; (4) citizen freedom and privacy fears, where despite increased human rights rhetoric, many people worry that personal information collection has gotten out of hand; and (5) international anarchy, where despite alleged enlightened global restraint and mutual respect, there appears to be a resurgence of "might makes right" behavior.

1

Global Information Overload

WITH REMOTE SATELLITE SURVEILLANCE, computers, the Internet, twenty-four-hour news coverage, social media, and cell phones, a dizzying flood of information bombards us with accelerating speed on a daily basis. We experience *information overload*, a term first popularized as a future trend back in 1970 by Alvin Toffler,[1] in which "the needs posed by information management exceed the capacities of the decisionmaking system"[2] because "we produce far more information than we can possibly manage, let alone absorb."[3] Today, "never in human history has more information been available to more people";[4] the amount of data we store doubles every eighteen months,[5] and Americans daily ingest five times as much data as they did twenty-five years ago.[6]

Compounding information overload is the dramatic recent escalation in digital data dependence. Government officials, business executives, and private citizens no longer see digital data access and reliability as a luxury but rather as a basic need and—within some societies—as a basic human right. In today's world, there is little tolerance of any form of downtime or delayed or interrupted access, with people incessantly demanding that any disruption be fixed immediately and exhibiting high anxiety or anger if it is not. Military digital data dependence has become especially tight, with digital technology in the U.S. military—consisting of fifteen thousand networks and seven million computing devices across hundreds of installations in dozens of countries employing over nine hundred thousand people—facilitating logistical support, command-and-control systems, real-time provision of intelligence, and remote operations.[7] Warfare is "no longer primarily a function of who puts the most capital, labor and technology on the battlefield, but of who has the best information about the battlefield."[8] Modern military combat now requires speedy and reliable data on remote targets and coordinated multifaceted strategy and tactics in the field, and operating blind—even with overwhelming force advantages—is a sure path to defeat. So accurate digital data access has become as critical to victory as military preparedness, troop strength, and advanced weaponry to succeed on the battlefield, and "information warfare is often cited as the *leitmotif* of early 21st century conflict."[9] One analyst quips, "If you want to shut down the free world, the way you would do it is not to send missiles over the Atlantic Ocean—you shut down their information systems and the free world will come to a screeching halt."[10] This psychological addition to digital data is thus highly dangerous and creates huge security vulnerabilities.

The global information explosion results from rapidly growing Internet and digital data technologies and recent advances in transportation and communication accelerating the pace of human interaction. Indirectly, the push toward "institutional transparency"—with legislation such as the Freedom of Information Act allowing outside parties to find out details of past government actions[11]—has accelerated information overload. This transparency emphasis, reinforced by the global spread of democracy, can sometimes affect even sensitive security matters, such as pressure to reveal details of arms control agreements,[12] with a 2016 White House report asserting that in law enforcement "measures that promote transparency and accountability" can particularly promote "trust and public safety in the community."[13] Such transparent data dissemination is often assumed to be universally desirable, leading to more accurate understanding of the world, easier crisis resolution, and possibly even alleviation of the security dilemma, in which minor disputes can escalate into major conflicts because of participants assuming the worst about each other's intentions.[14]

Contrasting Reactions to the Information Explosion

Major heated debates rage over whether the information explosion is a blessing or a curse. Supporters are constantly "heralding a new golden age of access and participation" and skeptics "bemoaning a new dark age of mediocrity and narcissism."[15] These opposite views reflect the reality that "we thrive on the information, and yet we can also choke on it."[16]

The Information Explosion as a Blessing

Technological optimists believe that the information explosion is an unambiguous blessing. They contend that as "information is moving faster and becoming more plentiful," "people everywhere are benefiting from this change."[17] To many observers, "in our age of instant information the benefits of speed and efficiency can seem unalloyed, their desirability beyond debate."[18] Claims of information access gains seem to know no bounds—for example, Bill Gates, the founder of Microsoft, asserts that "the information superhighway will bolster democracy, spread educational advantages to even the poorest kids, and usher in a world of 'low-friction, low-overhead capitalism . . . a shopper's heaven.'"[19] A related common sweeping claim is that "an almost-free world of information will bring equality of opportunity to all."[20] These alleged information explosion benefits cause the solution to every problem to revolve around seeking out more data.

In response to information overload concerns, information explosion advocates argue that improved filtering would easily manage any data excess encountered. Indeed, many of these advocates assert that "the information overload idea—that too much information causes dysfunction—is a myth" because "when choice sets become large or choice tasks complex relative to consumers' time or skill, consumers satisfice rather than optimize" and "do relatively well" in the process.[21] Thus faith exists that no matter how much data emerges, it will be processed properly.

This rosy logic, implying that more data are always better, rests on the following assumptions:

- Growing information access will stimulate the development of better information filters, allowing information consumers to find what they need efficiently.

- The more information is available to foreign security policy makers, the more accurate their perceptions are, the more balanced their understanding is, and the more confidence can emerge that resulting foreign policies will be sound.

- The more information is available to foreign security policy makers, the less likely an adversary would be able to use ambiguity, deception, and surprise against them (unless its back is to the wall in an asymmetrically disadvantaged position)

- The more information is available to foreign security policy makers, the greater power and influence they will have in international relations.

- The more information is available to foreign allies, the greater their abilities to help out with supportive actions.

- The more information is available to private citizens (from public and private sources), the greater the transparency of government action and the better input they will provide to foreign policy makers on foreign policy issues.

- The more information is available to everyone in the world, the more everyone will have an opportunity to advance because that access will open the door to learning and opportunities otherwise unavailable.

- As a result of the aforementioned assumptions, the quest for more information in the form of usable foreign defense intelligence ought to be one of the primary policy emphases for government security officials.

The Information Explosion as a Curse

In contrast, technological pessimists view the information explosion as a curse. They argue that the Information Age "has increased the difficulty of determining whether information is relevant or just random noise."[22] "Fake news" is proliferating—for example, during the 2016 American presidential election campaign, the mass public "was assaulted by imposters masquerading as reporters—they poisoned the conversation with lies on the left and on the right."[23] Having tons of information at one's disposal can create overconfidence even among experienced intelligence analysts in interpretative judgments,[24] especially problematic when dealing with culturally contrasting adversaries.[25] The common notion that "information is power" in this view may be "one of the great seductive myths of our time":

> When it comes to information, it turns out that one can have too much of a good thing. At a certain level of input, the law of diminishing returns takes effect; the glut of information no longer adds to our quality of life, but instead begins to cultivate stress, confusion, and even ignorance. Information overload threatens our ability to educate ourselves, and leaves us more vulnerable as consumers and less cohesive as a society.[26]

Information overload can become a truly overwhelming burden, making people feel "anxious and powerless," reducing both creativity and productivity.[27] Debilitating mass confusion can easily result from too much information and too little time to deal with it.[28] Furthermore, digital data proliferation can amplify security vulnerabilities: "In this new age, interconnectivity and dispersed computing power have significantly increased access to and dependence upon information, making the places it resides (such as databases, programs, and networks) more attractive targets" for adversaries.[29]

This gloomy logic, implying too much data is problematic, rests on the following premises:

- Human information processing capacity cannot keep up with the huge and accelerating increases in data access.

- People experience rising difficulty in distinguishing real news from fake news, regardless of the availability of multiple sources of information.

- Increasing digital data dependence increases vulnerability to disruption by adversaries.

- Deference to digital data may foster unfounded beliefs about numbers revealing objective truth; alternatively, contradictory information may trigger fatalistic beliefs that global security truths will never be discovered—"your truth is your truth," and "my truth is my truth."

- Greater global information access may increase undesired ambiguity, deception, and surprise in foreign security policies.

- Growing information access could raise foreign security policy makers' overconfidence in their conclusions or their misunderstandings of important issues, because of inconsistent evidence or the mass public's confusion, stress, and feeling of powerlessness to control their fate.

- Pervasive high foreign security policy maker expectations of sizable payoffs from greater information access may lead to widespread elite frustration, resentment, and disappointment.

- Growing information access for the mass public could lead to (1) confusion, stress, and a feeling of powerlessness to control its fate; (2) unrealistic expectations about success in government foreign security policy; or (3) unrealistic demands about changes when policies fail, causing frustration, resentment, disappointment, and authority distrust.

Comparative Pre-Internet-Age Retrospective

Analyzing the distinctiveness of the information explosion requires open exploration of whether it really has significantly changed data interpretation or strategic ambiguity, deception, and surprise. To justify this study's contention that information overload linked to the Internet age has indeed triggered a major transformation, a well-known earlier case involving strategic ambiguity, deception, and surprise—the 1973 Yom Kippur War—is briefly explored. Although this war is certainly not representative of all pre-Internet-age data interpretation or strategic manipulation, it can at least illustrate what has dramatically altered.

A key information handicap in the Yom Kippur War was Israeli uncertainty and surprise—in part due to a poor signal-to-noise ratio—about the timing of the Egyptian and Syrian attack, which was decided by these adversaries at the last minute.[30] Although the American Central Intelligence Agency (CIA) had received a credible report in late May 1973 that Egypt and Syria would initiate a war against Israel on October 6,[31] that alert was obscured by Egyptian deception, denial, and disinformation.[32] During the Internet age, despite vast improvements in surveillance technology, the proliferation of unreliable social media outlets, fake news, and ambiguous enemy messages would probably dramatically worsen the signal-to-noise ratio and the probability of accurate interpretation of what was about to transpire.

Another information handicap evident in the Yom Kippur War was the mistaken Israeli belief that the Russians would restrain an Arab attack on Israel out of respect for the existing calm detente between the United States and the Soviet Union.[33] When considering outsiders' interfering efforts, partly because the Cold War bipolar world was much simpler than the post–Cold War multipolar world, in today's Internet age there would probably be a far wider range of potential participants, including both state and nonstate players, using a far wider range of intervention techniques, including both real-world and digital means of intervening. The result would make the calculus of probable intervention much more difficult (for example, cyberattacks against Israel emanating from nonstate groups in the Arab world are now commonplace).[34]

A third information handicap was Israeli intelligence's underestimation of the enemy Egyptian and Syrian capabilities prior to the Yom Kippur War. This misperception was enhanced by ambiguity-enhancing Egyptian circulation of rumors about inadequate maintenance and lack of spare parts for their anti-aircraft missiles, and Egyptian concealment of added new weapon capabilities to their arsenal,[35] buttressed by the Israel Defense Forces' arrogant bias about Israeli air superiority.[36] During the Internet age, in which digital technology has readily facilitated the use of fake tanks on the battlefield and in which computerized weapons capability enhancements are much subtler and harder to spot through remote sensors than traditional improvements in weapons' mobility, lethality, or distance, the "fog of war" seems likely to be considerably denser.

This pre-Internet-age case suggests that the information explosion triggered a dramatic change not in form but rather in severity, with data interpretation much harder today and manipulation much easier than before. Nonetheless, certain constants are evident over time. Then as now, differing cultural values cause miscalculation: illustrating how "non-Western societies have often tended to undertake much greater risks in war than were considered rational or profitable by Western standards," the Israelis did not comprehend how "the Arabs were willing to risk military defeat, which is what occurred, in order to ameliorate their political position"[37] (the Egyptians succeeded in all of their manipulation goals).[38] Similarly, then as now, "decision makers who had in their possession all the necessary data failed to arrive at the correct conclusions."[39]

Big Data Analysis Promises and Perils

Given the growing importance of the Internet, a wealth of digital data has emerged, calling out for new forms of assessment. With traditional print (newspaper and magazine) and broadcast (radio and television) media, a relatively clear picture has emerged about how to interpret and evaluate their credibility and explicit or implicit bias. At the same time, however, because of low barriers to entry and to the growing ubiquitousness of cell phones and other mobile computing devices, the influence of Internet sources—particularly social media—has begun to rival and even eclipse that of traditional print and broadcast media. "With over 2 billion Internet users and the rise of social media, there is far more information than we can possibly process."[40] Internet-based news sources have sprouted up like weeds, with their number becoming so multitudinous and their management so rapidly changing that credibility, bias, and even basic accuracy become much harder to track. For many people around the world, their primary news organs are sources without substantial credibility monitoring, such as Facebook, Twitter, and YouTube. Those attuned to digital data often prefer short messages, pictures, and video clips over detailed textual analysis, reinforcing an emphasis on short-term memory, emotion, and simplified black-and-white judgment rather than logical, qualified, empirically based arguments.[41] The CBS news program *60 Minutes* reported that "people in general are quick to believe anything that is put in front of them in a format that is 'newsish,'" and that fake news has managed to trick both the educated and the uneducated.[42]

The growing availability of huge amounts of digital information resulting from extensive surveillance and sensors, combined with sophisticated means of data analysis undertaken by powerful data-crunching supercomputers, caused the *New York Times* in 2012 to declare that we have entered "a new epoch in human affairs—'the age of big data.'"[43] Computerized "big data analysis"—which "refers to things one can do at a large scale that cannot be done at a smaller one, to extract new insights or create new forms of value"[44]—can indeed facilitate breakthroughs in understanding of complex relationships previously incalculable by human experts, and has promised to make major leaps forward in human understanding of a wide variety of local, national, and global trends and their interrelationships. In finding optimal solutions to pressing and complex global problems, it works through "applying math to huge quantities of data in order to infer probabilities" about future patterns.[45] The vast information needed for big data analysis derives from "large, diverse, complex, longitudinal, and/or distributed datasets generated from instruments, sensors, Internet transactions, email, video, click streams, and/or all other digital sources available today and in the future."[46] The underlying premise behind the push for big data analysis is that "we can learn from a large body of information things that we could not comprehend when we used only smaller amounts,"[47] with complex data analysis software revealing "utterly unexpected correlations" yielding novel insights that change the way we think about the world.[48] The prospects for this kind of new understanding leading to constructive social engineering and improving needed services seem huge, at least on the surface: "when it comes to generating economic growth, providing public services, or fighting wars, those who can harness big data effectively will enjoy a significant edge over others."[49]

Big data analysis optimism is based on positive premises about technological progress. A common belief is that "modern science and technology truly have made it reasonable not only to expect to make decisions 'with confidence' but also to aspire to having information that is 'complete, true and up-to-date.'"[50] These expectations in turn rest on rosy assumptions that "necessity is the mother of invention"—no matter how challenging future data interpretation problems may become, or how sophisticated the resulting strategic manipulation of data may be, human ingenuity will find ways to overcome them, with big data analysis advances leading the way.

However, big data analysis is inescapably reliant on the validity and reliability of the information collected. The infamous longstanding GIGO pattern—garbage in, garbage out—still applies. While many analysts assume big data analysis is immune to bias because of the immense amount of information collected and the impartiality of data assessment algorithms, in reality even completely data-driven analysis may be imperfect because collected information may be incomplete, incorrect, or outdated.[51] When it comes to intelligence on adversaries, credibility and transparency are often inversely related. Big data analysis assumes that if data collection is comprehensive enough, any existing information distortions will be washed out by the overarching trends revealed through the preponderance of accurate valid and reliable data, but in many cases most of the data collected may be misleading. Even big data advocates admit that "when we increase the scale [of data collected] by orders of magnitude, we might have to give up clean, carefully curated data and tolerate some messiness,"[52] although they optimistically—and perhaps naïvely—argue that "with less error from sampling we can accept more measurement error."[53] Moreover, data analysis algorithms themselves "are not infallible" because "they rely on the imperfect inputs, logic, probability, and people who design them."[54] Such algorithms often rely on controversial assumptions about thresholds of deviation or significance used in isolating critical data anomalies. So big data mining of vast quantities of information can be a real blessing if and only if (1) sound means exist to determine information credibility; (2) data analysis algorithms can appropriately separate major anomalies from minor fluctuations; and (3) data consumers find ways to interpret emerging patterns properly and place them in an appropriate context. If any one of these three prerequisites is not fulfilled, then confident trust in big data analysis findings and recommendations can become dangerous. As a result, in practice recent applications of big data analysis have often resulted in "grievous mistakes."[55]

Despite these drawbacks and limitations, in the global security realm big data analysis has become highly popular. Within foreign intelligence, "in the past decade, agencies like the CIA and the NSA have institutionalized big data through the development of dedicated analytics units and research and development projects focusing on the analysis of online data."[56] However, as shown later in this chapter's security information unreliability analysis, because of widespread security data secrecy, especially on sensitive intelligence issues, information publicly available may be misleading, and information covertly collected on newly emerging threats may have uncertain credibility. Moreover, big data analysis systems are often eminently hackable by adversaries. In carefully executed global intelligence work, determining the accuracy and relevance of data collected can be a painstaking human-intensive task, not easily replicable through any fixed computer algorithm and or machine-learning system. Generally, "data analytics algorithms (primarily machine learning) are designed to solve problems where we have a prior understanding of the complexities in data sets," which is often not the case regarding foreign defense and intelligence data.[57] Moreover, big data analysis seems most useful when dealing with predictable environments characterized by continuous evolutionary change, not unpredictable settings evidencing the kinds of frequent discontinuous radical changes characterizing many important global security developments.[58] Thus, it is extremely difficult for big data analysis alone—with limited time-frame applicability and unexpected shock explanatory power—to reliably bridge significant foreign intelligence knowledge gaps, creating strategic uncertainty.[59]

Furthermore, even if most foreign intelligence collected were to be deemed credible, the results of big data analysis can be difficult to interpret properly. Often data available are "only a proxy for what you really want to know," and the logical leap made from one to the other can be highly dubious: usually big data analysis promotes decisions based on elements we know how to measure, frequently ignoring in interpretation critical intangible elements we do not yet know how to measure.[60] Determining such meaning can be challenging because big data analysis can ignore crucial elements of the foreign security policy context (commonly termed *domain knowledge*) or subtle clues or shades of meaning:[61]

> The core of military intelligence is to assess the capabilities, intentions and activities of the targets. This will not be accomplished simply by employing methodological processes that mine the wealth of data in search of patterns and trends that can provide insights into the target. Technology and technical competence can help but, in the new security environment marked by nuances, subtle political gesturing and signaling activities combined with the ability of the target to employ counter measures to circumvent the best efforts of the intelligence communities, there is no replacement for an experienced, well-trained analyst that can harness all available resources to produce informed assessments.[62]

When major cross-cultural differences are evident in foreign intelligence, it seems especially likely that big data analysis can inadvertently lead to significant misinterpretation. If the role of human judgment is marginalized, what particularly suffers is assessing enemy intentions,[63] whose understanding is so critical to foreign intelligence predictions and responses. So while relying on big data analysis is seductively attractive, even the best artificial intelligence or machine-learning algorithms can be severely limited or even misleading when used alone to address foreign security data's unique challenges. Thus in the end consistent avoidance of major interpretive errors within an information-overloaded strategic-manipulation-laden environment requires that big data analysis by the intelligence and defense communities incorporate significant human inputs and guidance to provide the crucial interpretation context.[64]

Mass Public Global Data Shock Concerns

Information overload has supplanted information scarcity as a widespread global concern,[65] and confronting too much information can be "one of the biggest irritations in modern life."[66] "Total noise" has created "the sensation of drowning and also of a loss of autonomy, of personal responsibility for being *informed*."[67] The information interpretation barriers seem so high that even when we are inundated with information on all sides of an issue, we can still feel ill-informed—"a barrage of data so often fails to tell us what we need to know."[68] These tendencies thwart any quest for clarity about what is actually transpiring.

Mass Public Annoyance with Information Overload

Despite some mixed opinion poll evidence,[69] citizens seem increasingly bothered, stressed, confused, and fearful about how to absorb, interpret, and respond to the information explosion.[70] Many people now even "feel overwhelmed by managing the most basic aspects of life,"[71] and constantly "complain of information overload—everyone is whiplashed by the changes."[72] Even among younger people, "the abundance of news and ubiquity of choice do not necessarily translate into a better news environment." In a major Associated Press survey, respondents "showed signs of news fatigue" and seemed "debilitated by information overload and unsatisfying news experiences," creating "a learned helplessness response," in which "the more overwhelmed or unsatisfied they were, the less effort they were willing to put in."[73] Proliferating information can increase citizens' cynicism about leadership performance and data subjectivity: some media skeptics wonder "whether truth even matters anymore—perhaps, they speculate, in the new information age reality is simply a matter of belief, not anything objective or verified."[74]

Mass Public Safety Deficit

As to citizens' sense of safety, the information explosion can inadvertently expand the range of perceived human security threats. Just as growing data access can increase potential victims' awareness of dangers, so can it cause threat sources to be able to "make use of the same technology to disrupt and counter intelligence collection efforts against them": "Information age technologies of the Internet and computers have provided new channels for international crime and terrorism" to wreak bottom-up havoc across societies.[75] The digital revolution has meant that more personal details of our finances, our behavioral tendencies, and even of our security measures have become more accessible to those who may wish to disrupt our lives, and that when such disruptions occur they are more widely publicized. Moreover, the triggering of mass public emotional responses—especially fear and national pride—on highly volatile issues can increase citizen vulnerability to the ravages of strategic manipulation under information overload, especially when as a primary manipulation target arousal of such passions heightens citizens' uncritical receptivity to messages confirming strong preconceived beliefs.

Even when people are comfortable with technology, the information explosion can complicate understanding of what is transpiring and reduce the ability of those most vulnerable to quickly adapt to changing conditions or to take appropriate steps to safeguard what is rightfully theirs. During times of imminent internal or external threat, central authorities' ability to guide citizens about appropriate safety responses could erode, because "in the midst of information overload, government communication is unlikely to reach citizens effectively."[76] This lack of clear guidance could lead to ensuing mass public chaos, along with its frustration about lack of control over its own destiny.

Mass Public Susceptibility to Negative Mass Media Influence

Given information overload, traditional print and broadcast mass media seem to have few incentives to provide balanced coverage of controversial issues or to strive to minimize bias injection likely to attract more attention. Instead, these outlets tend to push sensational stories that covertly promote whatever hidden agendas they (and their advertisers) support. Especially during fast-moving unexpected crises, the media can obfuscate as much as clarify the ongoing predicament. So even people who believe they have a clear global picture may be sadly misguided and out of touch with ongoing security realities.

Because Internet-based social media allow individuals to upload virtually anything they want, and because determining upload data quality is so difficult, information consumers often cannot distinguish fact from fiction and in some cases have given up trying to do so. While via social media many citizens gain a greater voice in socioeconomic disputes and more say in shaping state policies, "those who would manipulate the public for political gain or profit—be it corporations or the government—have more direct access to the public as well": for example, many websites pretending to be "independent informa-tion providers" are controlled by political groups.[77] As a result, interpretation distortions chances rise, as "mediating technologies make it more difficult to communicate intentions and easier to distort what the intentions may be."[78] For example, even with the possible injection of emojis and emoticons, communi-cating robustly true intent or subtle feeling nuances through e-mails and texts is virtually impossible. A double-layer interpretation challenge thus magnifies security risks, in which people have to filter out distortions both in the data it-self and in the social media technology conveying it. In the end, democratizing information access and news creation via social media can inadvertently leave people befuddled.

Mass Public Concern About Big Data Analysis

Global citizens are somewhat fearful of both the kind of information collection enabling big data analysis and its intrusion on their personal lives.[79] For exam-ple, "the broader public increasingly sees the U.S. government as having much too much technical capability for collecting, storing, and processing individual-level data, which in turn allows it unprecedented access into the everyday lives of millions of people—including U.S. citizens."[80] Big data profiling can be used to influence mass public sentiment (without citizens being aware of it) on key issues. The result in many democracies is that "the tension between the con-venience of data integration and fears of government overreach is palpable."[81]

The potential invasion of personal privacy, imposition of penalties or loss of innocence based on personal propensities, promotion of socially regressive policies, uneven access to benefits, and misuse for repressive purposes all are citizen concerns.[82] For example, even if specific names are withheld, some citi-zens consider concluding that people of a given type within a given area have certain tendencies to be a highly irritating violation of personal space. A major American government study concluded that "while big data can be used for great social good, it can also be used in ways that perpetrate social harms or render outcomes that have inequitable impacts."[83] Citizens rarely are convinced by common security justifications for personal data collection. Data gathered by trusted sources for legitimate purposes can be shared, sold, or stolen, placing it in the hands of those with no restrictions on its use. In many societies, "big data exacerbates the existing asymmetry of power between the state and the people."[84] There has been insufficient transparency and open discussion about the norms guiding collection targets, techniques, analysis, and dissemination relevant to citizens.

Information Interpretation Barriers

Humans have difficulty drawing correct meaning—whereby receiver understandings correspond with sender intentions—from incoming communication. The world has increasingly been "one of growing complexity, rising uncertainty, and diminishing capacity to anticipate and control outcomes,"[85] where increased information can foster misperception and manipulation.[86] As a result, the legacy of foreign security policy making is replete with examples of "human error and fallibility," which usually "did not stem from a dearth of information but rather from incorrect judgment and evaluation of available information"[87]—"the main shortcoming of our information culture is not the availability of 'facts'" but rather in determining "their quality and relevance" to those receiving the data.[88]

Within the global security realm, the core interconnected information interpretation barriers, depicted in Figure 1.1, are (1) data quantity or quality distortions associated with escalating information overload and security information unreliability (the stimulus); (2) receiver processing limitations associated with human cognitive frailty and organizational decision inflexibility (the response); and (3) system value heterogeneity associated with global cultural diversity and international political anarchy (the context). Together these create a nightmarish perfect storm for state and nonstate players in terms of the high strategic manipulation potential on foreign security issues (the outcome), including the ability to initiate it and to become a victim of it. So it is perhaps no surprise that foreign manipulation is so much more tacitly accepted in global compared to personal interactions: because of differing norms, "the exposure of one kind of manipulation apt to cause great discomfort in interpersonal affairs is not a major problem in international relations," in which manipulation initiators often seem quite willing "to incur high short-run costs in the expectation of receiving even larger long-run advantages."[89]

How Escalating Information Overload Can Impede Data Interpretation

Recent trends facilitate escalating information overload hindering data interpretation:

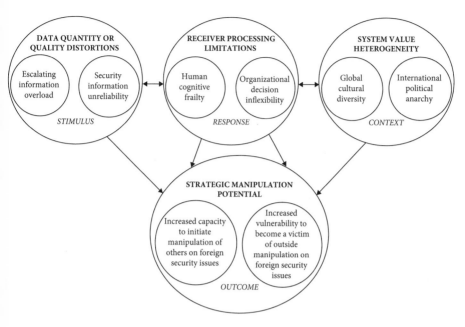

Figure 1.1. Information interpretation barriers

- Multiplicity of information sources on global predicaments, even in re-mote undercovered parts of the world
- Difficulty in determining credibility of original sources from which new information derives
- Increased "datafication" of otherwise immeasurable political, economic, and social phenomena[90]
- Increasing global desire by private citizens for more government trans-parency in foreign relations
- Greater global information dependence reflecting much greater global technology dependence
- Accelerating pace of change in global predicaments, facilitated by the speed of communication and information transmission
- Emergence of complex global threats cutting across national boundaries requiring broader profiling and modeling
- Globalization-induced needs for both states and nonstate groups to share information with others so as to jointly tackle common problems

- Internationally proliferating computing and communication devices and sensors, leading to mushrooming digital data generation

Because many now see data access as a basic survival need, slowing down these trends is difficult.

Information overload can obstruct data understanding in several ways, summarized in Figure 1.2. It can result in (1) many contradictory data sources exhibiting dubious credibility, impeding distinguishing believable news from deceptive news; (2) the constantly accelerating rate of change of data, impeding up-to-date interpretation of current trends; (3) worsening communication signal-to-noise ratios,[91] impeding distinguishing critical information from extraneous information; (4) difficulties learning from past data interpretation errors, because of low correspondence between accurate interpretation and successful outcomes; and (5) digital database vulnerabilities to hacking by disruptive parties that corrupt or distort information for their own nefarious purposes. For the first obstacle, given that news sources draw stories from other feeds capable of reporting "fake news," it becomes hard to determine systematically what to believe. Often "windfalls of accurate information are rejected as plants, since deception, as all actors are aware, is often based on the supply of seemingly accurate and verifiable data."[92] For the second obstacle, "the pace of technological change combined with the undisciplined inclination to assume more technology is better—techno creep—generally exacerbates the effects of data smog and information overload."[93] For the third obstacle, given cognitive limitations, "we can have trouble separating the trivial from the important, and all this information processing makes us tired,"[94] particularly problematic when dealing with low-probability, high-impact phenomena.[95] Security analysts regularly find that "it's possible to have too much threat data," obscuring the actual signals involved and causing security analysts to dismiss or ignore what is most critical;[96] even silence can become "unintentionally produced noise when it is interpreted by one of the participants as tacit but meaningful communication."[97] For the fourth obstacle, "recognizing successes and failures in information processing is sometimes tricky because of their disguised manifestations":

> When a failure in information processing does not produce high-cost outcomes because external circumstances intervene, significant failures may go unnoticed and the failure may even be mistaken for a success. Similarly, when a success in information processing is not utilized and policy and policy implementation consequently produce high-cost outcomes, the success may be mistaken for a

> **DUBIOUS SOURCE CREDIBILITY**
> Many contradictory data sources have exhibited dubious credibility, so it is hard to identify which is believable or deceptive.

> **RAPID NEWS TRANSFORMATION**
> The constantly accelerating rate of change of data makes it hard to keep up-to-date in one's interpretation of current trends.

> **POOR SIGNAL-TO-NOISE RATIO**
> Finding critical relevant information within a sea of extraneous irrelevant data is becoming more difficult.

> **DIFFICULTY CORRECTING ERRORS**
> Learning from past interpretation errors is hard because of mismatches between accurate interpretation and successful outcomes.

> **CYBER HACKING VULNERABILITY**
> Digital data are increasingly vulnerable to hacking by disrupters seeking to corrupt or distort information for their own purposes.

Figure 1.2. How escalating information overload can impede data interpretation

failure. In both cases feedback produces the wrong kind of learning because the inferences from outcomes are misleading.[98]

Concluding with the fifth obstacle, former director of national intelligence James Clapper worries about sinister operations that "change or manipulate electronic information to compromise its integrity, instead of simply deleting or disrupting access to it,"[99] and Admiral Michael Rogers, former director of the National Security Agency (NSA), fears the heavy security policy reliance on manipulated digital data—"what happens if the digital underpinning that we've all come to rely on is no longer believable?"[100] Vastly differing cultural lenses impede consensus in assessing data quality, relevance, credibility, and meaning.

How Security Information Unreliability Can Impede Data Interpretation

Security information unreliability can constitute a major interpretation blockage, as shown in Figure 1.3. With more foreign security data emanating from both public and private sources, often in precise quantitative form, many information consumers blithely presume that what they receive is both valid and reliable. This premise is based on an assumption that technological surveillance

HIGH BLIND TRUST IN SECURITY DATA

Increasing foreign security data emanate from both public and private sources, often in precise quantitative form.
Many information consumers blithely assume that what they receive is both valid and reliable.

HIGH FREQUENCY OF DEFECTIVE SECURITY DATA

Critical defense information is typically classified and secretive, and foreign security data are notoriously untrustworthy.
Much defense information is distorted or misleading, preventing definitive strategic understanding of what is transpiring.
Capable adversaries regularly inject untruthful information or omit vital information and adeptly change data distortions over time.
Crucial information on rapidly changing security predicaments usually contradictory and available only from few sources.
Cultural and value differences in foreign security data obfuscate signal meanings and require background context knowledge.

HIGH VIGILANCE NEEDED CONCERNING SECURITY DATA MANIPULATION

Because of widespread cross-national security data unreliability, the potential for foreign security manipulation is quite high.
This data manipulation potential is often downplayed in data analysis.
Erroneous foreign security data can be either benignly fabricated or maliciously falsified.
Intelligence analysts need to be hyperaware that the raw data relied on may be false, misleading, or deliberately distorted.
Private citizens need to be wary when exposed to foreign security data in deciding whether identified threats are real or exaggerated.

Figure 1.3. How security information unreliability can impede data interpretation

advances have over time increased the accuracy of data collected and disseminated.

However, high-quality defense information is typically classified and secretive, and foreign security data are notoriously untrustworthy. Not only within a very noisy environment can most of the data collected be distorted or completely erroneous (given that security establishments try to keep the true data secret), but also capable adversaries regularly inject untruthful information or omit vital information and adeptly change data distortions over time, making traditional means of coping with messy data relatively useless. Moreover, not only is crucial information on rapidly changing ongoing security predicaments usually available only from few sources, but also such information is often contradictory, making typical means of sound credibility and accuracy checking to ensure data validity and reliability unworkable. During wartime, famed Prussian military theorist Karl von Clausewitz argued long ago that data unreliability is especially high,[101] as typically many communication verification channels are blocked.[102] Furthermore, given cultural and value differences embedded in foreign security data, not only are vital signals difficult to understand in terms of their literal meaning, but also they require an incredible amount of idiosyncratic background context information to determine what is really being implied, making standard means of deducing intent relatively futile. Thus given that intelligence analysts are usually expected not only to predict enemy action but also to make high-risk decisions about how to respond effectively, they need to be hyperaware that the raw data relied on may be false, misleading, or deliberately distorted.[103] In parallel fashion, private citizens need to be especially wary when exposed to foreign security data in deciding whether identified threats are real or exaggerated.

Because information overload exacerbates security information unreliability, foreign data manipulation seems especially likely, requiring exceptional vigilance. However, this manipulation potential is typically downplayed in data collection.[104] Two primary types of foreign security information manipulation are common—benign fabrication and malicious falsification. Of these two manipulation forms, malicious falsification is globally considered more illegitimate and appears to be more likely to generate target resentment, anger, and retaliation.

Illustrating foreign manipulation through benign fabrication are three quite different data distortions. First, some national governments, lacking the ability or willingness to systematically collect accurate security data, give global data collection bodies just rough or distorted estimates masquerading as precise figures. In several Global South states, where often "whether you're looking at economic growth, hunger, or poverty, the numbers are frequently inaccurate":

> When it comes to economic growth in Africa, in some respects, we know as little as Livingstone did going up the Nile. Figures are often absent, of poor quality, or contradictory. To understand why, you have to look at the source: the various national statistics offices. They're charged with providing international organizations with data, but they're hampered by a chronic lack of funding, staffing, and the knowledge necessary for collecting the required information.[105]

Authoritative estimates of such countries' gross domestic product (GDP) vary widely, and "it's not just the quality of GDP figures that's a problem, their timeliness is too—procedures are so sluggish that national statistics offices often take years to release figures."[106] Second, even vital survival data can be fabricated: within two sub-Saharan African locales, significant data fabrication occurred in medical field research, deemed likely whenever "highly-stratified and compartmentalised research is conducted against a background of social and economic inequality, high unemployment, and poor labour laws."[107] Third, in the Global North, although the exact size of illicit transnational criminal transfers of arms, drugs, and humans via black market underground networks is impossible to quantify accurately, states regularly report to global bodies precise yet widely varying figures about their size, taken by recipients at face value.[108] For such inadvertent data distortion, "international organizations—along with journalists and policymakers—should be forthright about the shortcomings instead of blurring our view of the world with false certainties."[109]

Illustrating foreign manipulation through malicious falsification is the tendency of some states and nonstate groups, intent on purposeful deception to achieve strategic advantage (often due to their lack of military superiority over adversaries), to transmit to adversaries false military information masquerading as accurate security intelligence. This pattern was notably evident in Operation Bodyguard, the World War II plan for the Allied invasion of Normandy in June 1944. This was "one of the riskiest, highest-stakes military operations in history," requiring the landing of a relatively small initial force to face potentially over 100 German armored, infantry, and fortress divisions, and its success "depended on preventing the highly alert Germans from knowing where and when to mass their combat power to repel the attack."[110] Overall, this ruse—involving "counterintelligence and especially counterespionage, cryptology, aerial supremacy, and the art and craft of deception through double-agent operations and decoy forces and feints"—turned out to be "such a monumental and masterful counterintelligence and deception operation that even perfectly rational individuals and organizations might have struggled to divine the truth through its fog and misdirection."[111]

How Human Cognitive Frailty Can Impede Data Interpretation

Human cognitive frailty consistently impedes sound information interpretation. Key anthropology and social psychology theories, shown in Figure 1.4, shed light on these limitations, often due to "misconceptions based on subjective judgments of how the other nation ought to behave rather than objective assessments based on how it is behaving" and to "the predominance of preconceptions over facts."[112] Such misperceptions highlight deep-seated conscious and unconscious tolerance of bias.

THEORIES OF HUMAN COGNITIVE FRAILTY

Cognitive consistency/confirmation bias: carrying over past images into the present, fitting incoming information into preexisting beliefs
Cognitive conceit: overcorrecting for identified perceptual errors by engaging in distortion in the opposite direction
Black-and-white thinking: rejecting nuanced meanings in favor of a moral self-image and a diabolical enemy image
Intolerance of ambiguity/need for closure: artificially creating certainty about what confusing messages or data mean
Wishful thinking: viewing the world as one hopes it would be rather than as it really is
Worst-case analysis: reflecting paranoid pessimism seeking to find evidence supporting dire predictions
Ethnocentrism: feeling superior to those not part of one's group or nationality
Xenophobia: amplifying fear of stangers or foreigners
Self-fulfilling prophecy: involving a prediction that directly or indirectly causes itself to become true
Selective attention: ignoring incoming information that contradicts preexisting images
Projection: ascribing to others characteristics and beliefs associated with oneself
Contrast projection: attributing to enemies characteristics and beliefs opposite to one's own
Rationalization: inventing artificial justifications for behavior that objectively would not be acceptable
Cognitive bolstering: seeking out evidence enhancing the credibility of preexisting beliefs

SECURITY IMPLICATIONS

Deep-seated persistent proclivities to avoid objective information assessment allow the predominance of preconceptions over facts
Limited ability—no matter how bright, well-intentioned, or well-informed analysts are—to interpret incoming information
Human bias enters the picture when using cognitive shortcuts to deal with complex and uncertain information
Result can be overconfidence in current attitudes and behavior and inattention to signals about deficiences in them

Figure 1.4. How human cognitive frailty can impede data interpretation

Cognitive consistency, or *confirmation bias,* carries over past images into the present, fitting incoming data into preexisting beliefs, so that previous expectations control current perceptions.[113] In the closely related *cognitive conceit,* cognitive consistency awareness causes a boomerang effect, leading to overcorrection and perceptual distortion in the opposite direction.[114] *Black-and-white thinking* incorporates a moral self-image and a diabolical enemy image, interfering with nuanced interpretations of ongoing predicaments.[115] The closely associated *intolerance of ambiguity* or *need for closure,* especially affecting terrorist groups,[116] artificially creates certainty about data meaning by demanding definitive answers on salient subjects.[117] *Wishful thinking* views the world as one hopes it would be rather than as it really is,[118] focusing just on positive outcomes in which desires take precedence over expectations[119] and often reflecting overconfidence and complacency. *Worst-case analysis*, the opposite tendency, reflects potentially paranoid pessimism seeking out evidence supporting dire predictions: "alarmists can always find a credible expert or report warning that terrible things could happen,"[120] typifying security analysts because it sees threat identification as an essential justification for the state to exist.[121] Buttressing this tendency are *ethnocentrism*, feeling superior to those not part of one's group or nationality,[122] and *xenophobia*, amplified fear of strangers or foreigners.[123] *Self-fulfilling prophecy* entails predictions that directly or indirectly cause themselves to become true, often about undesired expectations.[124] *Selective attention* ignores incoming information contradicting preexisting images[125] or emanating from differing value systems: regarding enemy intentions, policy makers risk "ignoring costly signals and paying more attention to information that, though less costly, is more vivid."[126] *Projection* ascribes to others characteristics and beliefs associated with oneself.[127] If dealing with enemy images, especially involving scapegoating, *contrast projection* can occur, attributing to others characteristics and beliefs opposite to one's own.[128] *Rationalization* invents artificial justifications for behavior that objectively would not be acceptable.[129] Last, the closely related *cognitive bolstering* seeks out evidence to enhance the credibility of preexisting beliefs.[130]

These human cognitive frailty theories underscore the limited ability of people—no matter how bright, experienced, well-intentioned, or well-informed—to accurately interpret incoming information: "fallible and resolutely imperfect people are part of command and control systems,"[131] and removing them from the loop would risk losing the critical "domain knowledge" of context impossible to obtain otherwise. Bias contaminates judgment and leads to defective decisions. The human mind is poorly "wired" to cope with too much uncertainty and complexity.[132] People resort to cognitive shortcuts to make information processing easier, impeding balanced objective security assessments.[133] The result can often be overconfidence and selective inattention to signals about deficiencies.[134] Analysts' security data interpretation thus requires cognizance of their own limitations to prevent them from being overly hesitant, tentative, and unwilling to make judgments, or overly defensive and prone to further justify existing views.

How Organizational Decision Inflexibility Can Impede Data Interpretation

Organizational decision inflexibility—highlighting policy maker unwillingness or inability to adapt in a timely manner to changing circumstances—can play a major role in data misinterpretation. Key communication and decision-making theories, shown in Figure 1.5, explain these limitations. The identified forms of decision dysfunction reveal deep-seated resistance to change in organizational standard operating procedures.

Groupthink reflects the tendency of small cohesive decision-making groups to suppress contrary viewpoints and develop an artificial image of unanimity even when choosing highly controversial policy options.[135] Often "this problem of misinterpreting or ignoring relevant information can occur when policy-makers allow a healthy consensus to slip into a static mindset that discourages alternative policy approaches"; "a decision-making process driven by such a mindset will both ignore dissenting information and analysis and discourage professionals in the field from offering dissenting advice in the future."[136] *Bureaucratic inertia* has organizational units resisting innovation, keeping threat perceptions and standard operating procedures from appropriately adjusting to changing circumstances, favoring data supporting superiors' bias over data challenging this bias,[137] promoting "grooved thinking," and assuming that what will occur in the future is simply an extrapolation of what has occurred in the past.[138] The bureaucratic politics literature emphasizes how competing organizational interests and agency parochialism can reduce vital data sharing

---

THEORIES OF ORGANIZATIONAL DECISION INFLEXIBILITY

Groupthink: small cohesive decision-making groups suppressing contrary viewpoints and developing an artificial image of unanimity
Bureaucratic inertia: organizational units resisting innovation and adjustment to changing circumstances
Cognitive closure: prematurely resisting change in beliefs by a group that may be hierarchical or compartmentalized
Satisficing: failing to perform a comprehensive evaluation of options and instead accepting the first minimally acceptable option

---

SECURITY IMPLICATIONS

Ignoring or filtering information perceived to be threatening or irrelevant to one's organization's culture or perceived mission
Rewarding definitive judgments and discouraging indecisiveness, which causes parroting of supportive interpretations of superiors' preferences
Pressuring analysts for definitive positive judgments even when facing an uncertain situation with a poor signal-to-noise ratio
Balancing risks of underreaction (complacence) with those of overreaction (crying wolf)

Figure 1.5. How organizational decision inflexibility can impede data interpretation

and tilt information interpretations and resulting policy recommendations toward those highlighting a pivotal role for one's own agency.[139] Bureaucratic inertia may associate with *cognitive closure*, premature resistance to change in beliefs once chosen by a group,[140] often enhanced by structural hierarchy, centralization, and specialization,[141] or by excessive compartmentalization evident especially within intelligence agencies.[142] Finally, *satisficing*[143] has a decision-making group failing to perform a comprehensive search-and-evaluation of all possible options for dealing with a predicament, instead ceasing consideration of alternatives when thinking of the first minimally acceptable option, which for foreign security policy-making groups often turns out to be an alternative necessitating application of military force.

Organizational decision inflexibility thus can promote data misinterpretation, especially if there is a low signal-to-noise ratio, quite common in foreign intelligence agencies: "although the intelligence community picks up warnings and threats all the time, the vast majority of these are false or exaggerated," and "many of those that turn out to be true are vague—a threat may exist but there may be little or no information as to when or where it will materialize."[144] In such settings, "key players tend to ignore or filter information perceived to be threatening or irrelevant to their organizations' culture or perceived mission," and the result is perception and policy distortion.[145] If definitive black-and-white judgments are rewarded and hedging and indecisiveness are discouraged, analysts tend to parrot supportive interpretations of their superiors' preferences or promote continuation of status quo policies. In this setting, typically "the military commander or policy level official more than ever wishes a judgment of certainty from the intelligence system" and "may press the intelligence system to come to a positive judgment despite the inherent uncertainties in the situation, or on the other hand, demand a degree of 'proof' which is absolutely unobtainable."[146] From an organization-wide perspective, "in response to threat reports or warnings, organizations are consciously forced to balance risks of underreaction (complacence) with those of overreaction ('crying wolf'), both of which can do great damage to organizational credibility."[147]

How Global Cultural Diversity Can Impede Data Interpretation

Despite the homogenizing influences of globalization, global cultural diversity can persistently complicate interpretation of foreign security information. As shown in Figure 1.6, data interpretation varies widely across differing cultures, often causing misunderstanding and hostility. Moreover, within cultural set-

OBTUSENESS IN UNDERSTANDING OTHERS' PERSPECTIVES

Interpretation of data varies widely across differing cultures, causing misunderstanding and hostility.
Contrasting values may cause societies to find others' communication styles opaque and unintelligible.
Precipitating future wars may be antagonisms among groups espousing opposing cultural practices and belief systems.

DIFFICULTIES RECOGNIZING ONE'S OWN BIASES

Interpreting information accurately necessitates sensitivity to the biases introduced by one's own culture.
Distorting one's own thinking are widely held beliefs, practices, or cognitive styles of one's specific social environment.
Judging foreign situations and practices in terms of one's own culture, habits, morals, and customs contaminates analysis.

PROJECTING ONE'S OWN VALUES ONTO OTHERS

Projecting one's own way of thinking onto others is common in interpreting outside information.
Assuming that others will act in a certain way because that is how you would act under similar circumstances is an error.
Failing to understand that others perceive their national interests differently from the way we perceive those interests is a constant source of problems.

Figure 1.6. How global cultural diversity can impede data interpretation

tings, different interpretations emerge between people who are private citizens versus elite policy makers, living in democratic versus autocratic political regimes, facing crisis versus routine challenges, and dealing with political-military versus socioeconomic issues. The high significance of cultural divides revealing contrasting values, whereby each group finds others' communication styles somewhat opaque and unintelligible, has caused some analysts to believe that a key trigger of future wars will be antagonisms among groups of people espousing opposing cultural practices and belief systems.[148]

Beyond the need to understand target cultures, accurate information interpretation necessitates greater sensitivity to biases introduced by one's own culture. Such cultural biases "are constraints on one's thinking, acquired during maturation from widely held beliefs, practices or cognitive styles that characterise one's specific social environment."[149] When one collectively judges others according to the values, morals, and customs possessed by one's own society, strategic thinking goes astray.[150]

Even when information consumers are aware of their own cultural biases, sometimes they still project their own way of thinking onto others in interpreting outside communication, constituting a real impediment to cross-cultural understanding. As a result, for example, American intelligence analysts frequently err when they engage in "filling gaps in the analyst's own knowledge by assuming that the other side is likely to act in a certain way because that is how the US would act under similar circumstances"—"people in other cultures *do not* think the way we do," and "failure to understand that others perceive their national interests differently from the way we perceive those interests is a constant source of problems in intelligence analysis."[151]

Meaningful data interpretation across cultures thus entails a complex set of skills. This skill set requires "an understanding of the attitudes and disciplines of potential adversaries as well as their capabilities, their history, their culture and their biases."[152] The specific kind of foreign cultural understanding for interpreting communication accurately and engaging in outside manipulation requires that one "must recognize the target's perceptual context to know what (false) pictures of the world will appear plausible," influenced by "history, culture, bureaucratic preferences, and the general economic and political milieu."[153] One example of a culture-based intelligence question directly related to accurate signal interpretation under information overload is "to what extent does the resurgence of long-dormant ethnic sentiments stem from a reaction against the cultural homogenization that accompanies modern mass communications systems?"[154] Such a question illustrates that a key prerequisite to understanding a foreign culture is comprehension of to what extent that culture seeks to emulate or resist the influences of other cultures, and to what extent the culture feels under siege and endangered versus secure and thriving. In the end, deep understanding of foreign culture cannot always be achieved alone or from a distance: consulting outside experts, and especially finding ways to listen directly to the perspectives of those who are now or have been in the recent past members of target societies, appears absolutely crucial to avoid "what Admiral David Jeremiah called the 'everybody-thinks-like-us mindset' when making significant judgments that depend upon knowledge of a foreign culture."[155]

How International Political Anarchy Can Impede Data Interpretation

International political anarchy, reflecting the absence of overarching common norms and meaningful authority structures worldwide, fosters a kind of every-state-for-itself mentality. In ways shown in Figure 1.7, such anarchy can hinder

---

INTERFERENCE WITH COMMON INTERPRETIVE NORM DEVELOPMENT

Anarchy interferes with developing a common set of global political norms facilitating communication and the search for meaning.
Anarchy creates fatalistic expectation that efforts to jointly or fully understand others' perspectives are doomed to failure.
Anarchy allows unruly parties using covert and hard-to-interpret signals to move easily across vulnerable targets.

VAGUE AND DYSFUNCTIONAL GLOBAL RULES OF THE GAME

There is little widespread understanding of—or compliance with—the set of anarchic rules of the game in international relations.
Each global party seems free to behave and interpret communication according to its own idiosyncratic premises.
Core powers cannot set by themselves the rules of the game, partly because of their lack of widespread perceived legitimacy.

DEGRADATION OF INTERNATIONAL THREAT COMMUNICATION

Targets may misunderstand the terms of a threat and thus may unknowingly violate them and cause the threat to be carried out.
The fluid, covert, multifaceted, and dispersed nature of threat sources clouds threat credibility.
Demonstrable, quantifiable, and clear threats are being replaced by fuzzy, fragmented, unquantifiable, and invisible threats.

Figure 1.7. How international political anarchy can impede data interpretation

data interpretation through obstructing the emergence of common global political norms that could facilitate communication and cross-national conflict resolution. In the end, anarchy can create a fatalistic expectation that efforts to act jointly or fully understand others' perspectives are doomed to failure.

Under anarchy, unruly parties can move easily across possible targets, choosing to operate in those with the most distracted, inept, or corruptible authority structures, or in those most naïve or baffled by the changing global threats. Such disruptive forces, using covert and hard-to-interpret signals, can then thrive and expand into effective transnational operations because their ability to overcome the rigidities of sovereignty, take advantage of anarchy, and mirror a kind of globalized efficiency may be far greater than that of most status-quo-supporting Western states.

There appears to be little understanding of—or compliance with—the set of anarchic rules of the game in international relations.[156] In the absence of a uniform and universal rule set consistently voiced and followed, each party seems free to behave and interpret communication according to its own idiosyncratic premises. Unlike in the past, core powers cannot set by themselves the rules of the game, at least in part because of their lack of universally recognized global legitimacy, which can increase their vulnerability to disruption.[157] The increasing popularity of moral relativism, with a premium placed on nonjudgmental tolerance of contrasting beliefs, can cause any thrust promoting a more coherent set of interpretation norms or rules of the game—especially by the West—to run the risk of being identified with the most virulent forms of cultural imperialism. In this way of thinking, to establish more universal rules seems akin to an anti-democratic squashing of everyone's ability to experience independent empowerment. For disenfranchised states and nonstate groups, the very notion of rules of the game in today's world is reminiscent of an era in which they sacrificed autonomy in foreign policy for what they perceived to be a quite arbitrary world order.

Moreover, for many disadvantaged parties unable to move up the global hierarchy, violating the rules of the game and thriving on misunderstanding can seem to be a way to escape from a stifling and humiliating status quo, a system whose premises they feel powerless to influence.[158] Those who do not want to play by the rules, including rogue states, terrorist groups, and criminal organizations, know that today it is difficult for major powers to exert effective long-run pressure on them, and indeed much of these noncompliant parties' status appears to derive from their ability to misinterpret or thwart flagrantly the major powers' rules of the game and to get away with it without suffering devastating consequences.

Within a global anarchic political system fraught with frustration and misunderstanding, when an instigator issues a threat, the target may not completely comprehend the terms of the threat and may therefore unknowingly violate them and cause the threat to be carried out. Even after being attacked, the target may not engage exhibit compliance in terms of the desired changes in behavior because its perception of what is expected is so ambiguous. Because disruption initiators are so fluid, covert, multifaceted, and dispersed, targets may remain uncertain about whether a threat will be carried out, and even after it is carried out the primary purpose and exact origin of dangers may remain obscure. Within such a degraded communication system, "demonstrable, quantifiable, and clear threats of global war are fewer; whereas fuzzy, fragmented, and less quantifiable—hence, less 'visible'—threats are legion."[159]

2 Global Strategic Manipulation

IN RECENT YEARS, strategic manipulation techniques have become more sophisticated, making it harder to discern whether one is being manipulated or whether information is clear, true, and accurate. Ambiguity, deception, and surprise are primary manipulative components of political leaders' strategic toolkits, and heads of state and leaders of nonstate groups do not shy away from using them. When aimed at adversaries, such approaches can "exploit vulnerabilities, capitalizing on hubris, complacency and self-delusion," and they can prevent accurate assessment of capabilities and intentions so as to hinder appropriate action.[1] Although sometimes "strong, confident nations like the United States lack the natural incentive to employ surprise, denial and deception," and may dismiss them as "weapons of the weak," such strategies have proven widely useful.[2]

Despite involving subjective manipulation, strategic ambiguity, strategic deception, and strategic surprise have widely accepted meanings. *Strategic manipulation* is projecting desired images consciously[3] for security gains in high-level broad policy formulation instead of low-level tactical implementation details, and as such, it encompasses the trio of strategic ambiguity, strategic manipulation, and strategic surprise.[4] *Strategic ambiguity* is the promotion of a "hazy middle ground" where "the information we need to make sense of an experience seems to be missing, too complex, or contradictory."[5] Reflecting either vague imprecise information open to multiple interpretations or seemingly conflicting statements and actions, ambiguity applies to communication content or to perpetrator and victim identities or motivations: the same message delivered in differing ways to differing audiences may be interpreted quite

33

differently.[6] Determining when such ambiguity is intentional versus accidental is often quite difficult. *Strategic deception* is a "deliberate misrepresentation of reality done in order to gain competitive advantage."[7] This includes both classic deception, inducing new information about something that is not true, and denial, blocking access to existing vital information,[8] as well as deception by commission (intentionally injecting false data) and by omission (intentionally leaving out crucial data). *Strategic surprise* is knowingly triggering an important, sudden, unanticipated security-transforming event, usually against a target that would have very much wanted to know about the development in advance so as to be able to forestall it or at least prepare for it. Specifically, "surprise is a strategic discontinuity, a startling seismic shock" that potentially "upends best laid plans, unbalances a comfortable posture," and "causes psychological dislocation and at least temporary paralysis."[9]

Under information overload, global security unpredictability is the primary consequence of widespread use of strategic ambiguity, deception, and surprise. Usually, being seen as unpredictable is considered normatively undesirable in international relations, assumed by observers to decrease chances of meaningful global stability and peace by disrupting smooth global diplomacy and cross-national treaties and commitments. Nonetheless, a strategic initiator promoting international images of unpredictability may gain advantages over adversaries, leaving enemies highly vulnerable because such images "can prevent an aggressor from guarding against a particular response."[10] The ability to appear unpredictable appears easier in today's world than in the past because of the global pervasiveness of ready access to huge amounts of contradictory information without simultaneously having ready access to tried-and-true methods for determining data accuracy, credibility, validity, or reliability.

Linking Information Overload to Ambiguity, Deception, and Surprise

Although typically discussions of information overload are quite separate from those of strategic ambiguity, deception, and surprise, and these three manipulation techniques are usually dealt with quite separately from each other, in reality all are tightly interrelated and require integrated analysis. Indeed, the interaction among these three manipulation techniques is so tight that often it is impossible to isolate cleanly the workings of one from those of the others. Figure 2.1 depicts a common but not universal multistage sequence of this interaction, whereby growing information overload leads to strategic ambiguity use, in turn facilitating strategic deception use and ultimately creating opportunities for strategic surprise.

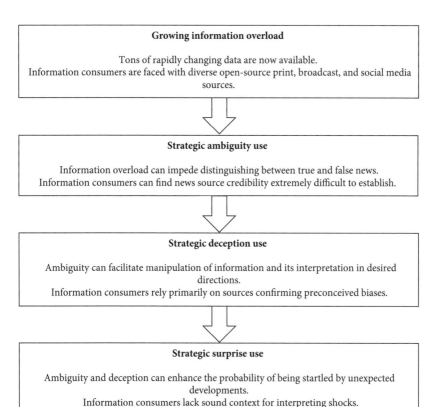

Figure 2.1. Linking information overload to ambiguity, deception, and surprise

Special linkages characterize information overload, ambiguity, deception, and surprise. First, growing information overload—involving tons of rapidly changing data—can promote strategic ambiguity because data oversaturation reduces information consumers' capability and confidence to distinguish between true and false news. The considerable noise present in today's global information overload degrades communication, whose "randomness, irrelevance, distractiveness, or lack of message" makes it "so ambiguous that it is difficult to decode."[11] Second, strategic ambiguity can promote strategic deception when ambiguity facilitates manipulation of information and its interpretation in desired directions to information consumers relying primarily on sources confirming their preconceived biases. Because ambiguity makes it difficult for recipients to determine the correct interpretation of signals, ambiguity is "almost essential in deceptive communication."[12] Third, strategic deception can promote strategic surprise, as both ambiguity and deception enhance the probability of being startled by developments for which there is no concerted anticipation, with data consumers lacking a sound context or necessary time to investigate or interpret shocks. Often called "the hand-maiden of surprise,"[13] deception is usually "critical to achieving surprise," for when combined, "surprise and deception produce a synergy, significantly increasing the chances of success."[14] Indeed, "surprise is a decisive factor in war, and it is by means of deception that one principally achieves this condition."[15] So strategic ambiguity, deception, and surprise can be mutually reinforcing when undertaken by the same initiator against the same target, and thus, potential targets' failure to think creatively and comprehensively about how strategic ambiguity, deception, and surprise could be combined can be dangerous.

However, despite these interconnections, information overload and strategic ambiguity, deception, and surprise are not invariably intertwined in this way. For example, rather than ambiguity causing deception, awareness of an adversary's deception could cause future interpretive ambiguity,[16] and rather than deception causing surprise, an unexpected shock could cause future information distortion to work better because of lower target confidence in its previous interpretive system. Moreover, strategic initiators do not always calculate in advance the exact desired pattern among ambiguity, deception, and surprise, and instead their plans may evolve as security predicaments progress. So sometimes advanced intelligence about an adversary's intentions may be insufficient to provide useful warning about impending manipulation, and sometimes special circumstances heighten the ambiguity-deception-surprise sequence's unpredictability. Furthermore, as noted earlier, information overload is enhanced in facilitating the frequency and success of strategic ambiguity, deception, and surprise by security information unreliability, human cognitive frailty, organizational decision inflexibility, global cultural diversity, and international political anarchy.

How Information Overload Can Transform Strategic Manipulation

The information explosion appears to have dramatically accelerated the ease and broadened the cross-national application methods of strategic ambiguity, deception, and surprise, if not fundamentally changing the nature of strategic manipulation. Figure 2.2 summarizes these wide-ranging changes.

SECURITY IMPACT ON TARGETS

Strategic ambiguity
Through greater reliance on too much contradictory data rather than on too little data and through the accelerating pace of data transformation, information overload can help strategic ambiguity worsen the signal-to-noise ratio and lower people's ability to perceive genuine new developments.

Strategic deception
By inducing people to believe something that is not true rather than on blocking people's access to vital data (denial), information overload can help strategic deception facilitate distraction, misdirecting people's attention and causing them to waste resources on unimportant purposes.

Strategic surprise
When rapid discontinuous shifts occur in complex interconnected data trends rather than simple gradual linear evolutionary change, information overload can help strategic surprise catch people off-guard, vulnerable, and unprepared, increasing the difficulty of accurate predictions of global security perils.

TARGETS' SECURITY RESPONSE

Strategic ambiguity
Through increased fatalistic pessimism by security policy makers about their inability to draw definitively accurate data interpretations, information overload can help strategic ambiguity enhance possibilities for significant long-term target decision-making paralysis.

Strategic deception
By preventing data consumers from discerning the difference between what is false or distorted and what is accurate and true, information overload can help strategic deception impede foreign signal understanding and successful defensive target countermeasures.

Strategic surprise
When causing the traditional defensive response—pouring more resources into intelligence collection and analysis—to be even more counterproductive, information overload can help strategic surprise interfere even more with political leaders' appropriate reaction to early warning signals.

Figure 2.2. How information overload can transform strategic manipulation

Strategic Ambiguity Transformation

Information overload enhances the potential for initiating strategic ambiguity and reduces the likelihood of successfully defending against it. Ambiguity is often "aggravated by an *excess* of data": for example, in attack warning, "there is the problem of 'noise' and deception," and in operational combat evaluation, "there is the problem of overload from the high volume of finished analyses, battlefield statistics, reports, bulletins, reconnaissance, and communications intercepts flowing upward through multiple channels at a rate exceeding the capacity of officials to absorb or scrutinize them judiciously."[17] When interpreting a flood of ambiguous information, personal or professional bias can enter the picture: "When the problem is an environment that lacks clarity, an overload of conflicting data, and a lack of time for rigorous assessment of sources and validity, ambiguity abets instinct and allows intuition to drive analysis"; "when a welter of fragmentary evidence offers support to various interpretations, ambiguity is exploited by wishfulness—the greater the ambiguity, the greater the impact of preconceptions."[18]

Under information overload, ambiguity can dramatically increase the time it takes for coherent decision making, especially if consensus is desired. So "in organizations that deal continually with potentially disruptive and threatening situations that involve a high degree of uncertainty, ambiguity, and complexity, such as organizations dealing with national security and foreign affairs, the top-level officials may adopt a wait-and-see attitude toward information about threats."[19] Foreign policy makers cannot usually delay data interpretation until uncertainty evaporates, for "in warning, delay can be fatal."[20] Especially in time-pressured crisis situations, "it is the job of decision makers to decide, they cannot react to ambiguity by deferring judgment."[21]

Information overload can affect strategic ambiguity in many ways. Information overload dramatically increases the proportion of ambiguity cases incorporating contradictions, increasing substantially the quantity but not the quality of available data and causing the signal-to-noise ratio and interpretive predicament to worsen substantially. Because of the accelerating pace of rapidly transforming data received, information consumers may find it more difficult to perceive genuinely new developments: "Even normally perceptive and imaginative individuals seem to have lapses in which they are unable to perceive that something new is occurring or that things have changed"—"they may be reluctant to accept evidence which in other circumstances, particularly when there is adequate precedent, they would accept without question."[22] Information overload tends to increase information consumers' fatalistic acceptance of foreign strategic ambiguity, rather than energetic searching for remedies, because of their pessimism that definitively accurate meanings can ever be found. Despite the understandability of such a gloomy response, it is not the most prudent response to ambiguity, as other viable, more promising options are available.

Strategic Deception Transformation

Information overload facilitates strategic deception, which "is a very powerful tool, going back to Adam and Eve"[23] and "is one of the oldest and most effective weapons of warfare":[24]

The success of any deception campaign rests on shaping the target's perception. The Information Age provides all the tools necessary to create a coherent comprehensive deception plan based on a sound knowledge of the enemy's perceptions and biases. The necessary transmission channels needed to shape that perception and create an alternate reality commensurate with the deceiver's objective exist in abundance. So too do sufficient feedback channels to enable continual adaptation of the deception plan all within in a shorter time cycle.[25]

Rising globalization and eroding national sovereignty help deceivers by opening many different dimensions of targets to outside inspection.[26]

The information explosion alters strategic deception in key ways. Because of greater data access, deception involving strategic denial—intentional blockage of targets' access to vital information—seems likely to decrease precipitously. Although in the few internationally closed societies—such as North Korea—this denial option is still possible, throughout most of the rest of the world the digital data revolution has made it difficult to completely close off information access in any sector, including defense. In addition, deception through commission (injecting false data) seems likely to become more common than deception by omission (leaving out crucial data), especially when the false data reinforce existing target beliefs and make concealment effective largely only in the short term because it is now easier to penetrate seemingly secure digital databases. On the other hand, at the same time it should be even easier for deception initiators to misdirect targets' attention away from vital security priorities, cause targets to waste resources on unimportant purposes, and catch targets off-guard and unprepared for imminent threats or crises. Although in theory greater information access could reduce opportunities for effective deception because of greater awareness on the part of targets of contradictory data challenging deceptively projected data, in practice this deception success reduction is rare because of the inability of data consumers within deception targets to find meaningful methods to evaluate definitively the comparative credibility of deceptive versus contradictory information.

Strategic Surprise Transformation

Information overload has raised the chances of strategic surprise. Although such surprise has occurred throughout history, it has been especially visible within the last quarter century:

One might usefully call the past dozen years "the age of surprises." The US government has been surprised by the end of the Warsaw Pact, the disintegration of the Soviet Union, the Iraq invasion of Kuwait and the ensuing Persian Gulf War, the Asian Financial Crisis, the Indian and Pakistani nuclear detonations, and now the events of September 11, 2001. There is no reason to think the age of surprises is over, and there are many reasons to think we are still at its beginning.[27]

Although "some strategists expect that advances in information technology will greatly diminish if not altogether obliterate" unexpected security shocks, "even with the best equipment in the world, militaries around the world frequently have been surprised by their adversaries."[28] Intelligence analysts trying to forestall strategic surprise "find it difficult to contend with large quantities of information, and this difficulty affects their estimates" in negative ways.[29] Generally, the greater the quantity of information collected, the more it is likely to contain noise, and when "information contains a large proportion of noise, the risks of another intelligence failure leading to surprise may even increase."[30] Unlike strategic ambiguity, whereby targets have at least a shot at correctly reading the tea leaves, sophisticated strategic surprise minimizes the target's potential to guess correctly what is about to happen. Although it is common to assume that the intelligence community fails to anticipate surprise attacks because of "a lack of relevant information," in contrast "in most cases the victim possesses an abundance of information indicating the imminence of the attack"; as exemplified by the warning about the Japanese attack on Pearl Harbor on December 7, 1941, the strategic surprise problem is often that within the sea of incoming data there is limited ability to distinguish the true signal from the extraneous noise.[31] American foreign policy makers seem particularly vulnerable to information-overload-induced strategic surprise: given vast global obligations, "issues accorded the highest priority by senior officials in Washington—who face huge burdens and competing demands—may overwhelm their time and attention, making it virtually impossible for them to pay attention to other issues being reported by professional analysts."[32]

Information overload also exacerbates the disruptive security impacts of strategic surprise. Accurate predictions of security perils could become even harder because of special problems spotting discontinuous shifts in existing trends, and so information overload may increase intelligence analysts' tentativeness and hesitation when encountering sudden major changes linked to strategic surprise, transformations that they cannot readily assimilate into their existing frames of reference. A dual danger exists here of oversensitivity, whereby every minor shift is inappropriately assessed as a major system shock, or undersensitivity, whereby even major shifts are treated as normal expected evolutionary changes not deserving any form of special attention or scrutiny. Establishing widely accepted and durable thresholds for unexpected major change seems extremely difficult given uncertainties surrounding data interpretation and fluctuation in acceptable global behavior norms. So today it may be harder to attain the gold standard for surprise preparation—"incorporating nimble and flexible modes of assessment not contaminated by preconceived biases, so as to avoid surprise, promote restraint, counter them effectively when restraint fails, and minimize losses."[33]

Last, information overload hinders post surprise target responses. The traditional post surprise defensive reaction—pouring more resources into intelligence collection and analysis—seems counterproductive because analysts are already overloaded with data of indeterminate relevance. The prospects for these postsurprise responses improving in the future appears to be low because of roadblocks impeding (1) senior officials absorbing and using field professionals' data and analysis and (2) political leaders reacting appropriately to threat warning signals.

Strategic Manipulation Ends

Strategic ambiguity, deception, and surprise are undertaken with explicit objectives in the minds of initiators, and similar such manipulation triggers explicit protective aims among intended targets to minimize its impact. While each target wants to maintain control of its society and make prudent choices, initiators want to wrest control out of its hands and induce it to distort what it decides to do. The contrasting offensive-defensive goals are summarized in Figure 2.3.

INITIATOR PREFERENCES	TARGET PREFERENCES
• Make the enemy quite certain, very decisive, and wrong in terms of its choice among perceived plausible alternatives.	• Deter and limit damage by identifying in advance *when*, *where*, and *how* an adversary can initiate manipulation.
• Cause adversaries to be confused, to waste time and resources, and to reveal their strengths, dispositions, and intentions.	• Enhance vigilance, defensive preparedness, and contingency planning, including preemptive actions to manage threats.
• Maximize outsiders' view that any preparations for manipulation are legitimate defensive ones rather than illegitimate offensive ones.	• Extend the lead time of strategic warning so targets have greater opportunities to contemplate the best ways to thwart foreign efforts.
• Make the target of manipulation behave in the desired way, placing it informally under the initiator's control.	• Increase the uncertainty of strategic initiators about the outcome of the manipulation efforts.

Figure 2.3. Offensive manipulation versus defensive response goals

Offensive Ends for Manipulation Initiators

Offensive strategic manipulation has several key goals. Most broadly, its objective is "to make the enemy quite certain, very decisive and wrong" in its choice of policies[34] so as "to make the victim act in the way we desire."[35] Such manipulation needs to confuse adversaries, causing them to waste and misallocate time and resources and to inadvertently reveal their strengths, dispositions, and intentions.[36] An underlying aim is to maximize outsiders' view that any manipulation preparations are legitimately defensive ones rather than illegitimately offensive.[37]

Regarding the specific goals of strategic ambiguity, whether using vague imprecise or directly conflicting data, ideally "the sending of ambiguous signals in a noisy environment allows the actor to leave the other in doubt, to influence the final impact of the signals after the initial reactions to them have been observed, and thereby to gain greater control of the images others have of him."[38] Ambiguity can facilitate for initiators plausible deniability of actions that could have negative consequences or even greater perceived legitimacy: especially within Western societies, "the ambiguity of security worked well to legitimize interventions," by which "it remained unclear whose security was pursued, either the local populations' one or the Western societies' one."[39] Such ambiguity is designed to make accurate interpretation of incoming information extremely difficult, for "when ambiguity and uncertainty increase, the information becomes more open to competitive interpretations and choosing among them becomes more complicated."[40]

Regarding the specific goals of strategic deception, the ideal aim is to promote "in the adversary a state of mind which will be conducive to exploitation by the receiver."[41] The idea is to create doubt in targets' minds about the extent to which the information received can be fully trusted as representing initiators' actual policies. Tactically, "deception can be cheap," for "little investment in men and material is necessary, and the return is very high."[42] As with ambiguity, a deception initiator must understand how targets would interpret incoming signals; with deception there is an added need to determine in advance what kinds of intentionally distorted or erroneous signals targets would consider most and least credible. Although many analysts see deception as a strategy for those disadvantaged in asymmetric confrontations, in many circumstances, it would seem that "more powerful military establishments must make a conscious effort systematically to incorporate deception into their military thinking."[43]

Regarding the specific goals of strategic surprise, initiators' ideal objective is, through catching each target unprepared, to put themselves in a more advantageous position to obtain what they want from it and leave it unable to fathom the exact nature of what is transpiring. As with strategic deception, initiators need to understand a target's modes of signal interpretation and signal credibility determination, but with surprise they also must be able to predict with some accuracy the target's reaction to unexpected shock. While some observers think that the opportunities and aims for strategic surprise have been forced to become narrower because of surveillance advances, in reality catching targets off-guard still occurs frequently with massive disruptive security impacts.

Defensive Ends for Manipulation Targets

Targets' defensive responses to strategic manipulation by hostile parties have aims diametrically opposed to those of deception initiators. Most fundamentally, "the goal is to deter and limit damage by identifying in advance *when*, *where*, and *how* a declared or potential adversary" can launch aggressive manipulative moves.[44] More narrowly, "the goal is to assist policy decisions on defensive preparedness and contingency planning, including preemptive actions, to manage the risks of potential threats."[45] As to decision speed, a key defensive objective is to "extend the lead time of strategic warning" so that targets have greater opportunities to contemplate the best ways to thwart the foreign effort.[46] Most subtly, aside from increased vigilance, target defenses can aim to increase strategic initiators' uncertainty about the outcome of their influence efforts.

The defensive ends for targets responding specifically to strategic ambiguity focus on creating clarity out of confusion. Because strategic ambiguity hopes to put target intelligence analysts in a bind with regard to identifying subtle communicating nuances, target intelligence analysts' primary role is "to extract certainty from uncertainty and to facilitate coherent decision in an incoherent environment"; this job is challenging because to the extent that target intelligence analysts reduce uncertainty by connecting the dots within ambiguous data, they "risk oversimplifying reality and desensitizing the consumers of intelligence to the dangers that lurk within the ambiguities," and to the extent that they do not resolve existing ambiguities, they "risk being dismissed by annoyed consumers who see them as not having done their job."[47] Moreover, target confusion is often enhanced about what message is being conveyed because when "the signals are ambiguous and the communications system noisy,"[48] the chances that targets' preconceived images will contaminate their interpretation of incoming information are increased.

The defensive ends for targets responding specifically to strategic deception include sharpening their ability to identify and differentiate key signals within incoming communication, discerning differences between what is true and untrue and what is important and unimportant. Without this defensive effort, "as a result of the great difficulties involved in differentiating between 'signals' and noise in strategic warning, both valid and invalid information must be treated on a similar basis—in effect, all that exists is noise, not signals."[49] Furthermore, the most ambitious goals for deception targets include causing initiators to fall into traps, disclose vital information, and become confused about outcomes. Getting a deception initiator off its game, so that it loses confidence that it is in control of the situation and that its manipulation is working, is the ultimate target objective.

The defensive ends for targets responding specifically to strategic surprise include trying to minimize its traumatic impact: for designated targets, surprise can be "sudden, stunning, traumatic, and humiliating," catching "the victim at his weakest, exposing and exploiting his failings."[50] Potential targets need to comprehensively gather what predictive indicators they can of adversary activity and prepare for postsurprise recovery and resiliency, because "strategic surprise occurs to the degree that the victim does not appreciate whether, when, where, or how the adversary will strike."[51] Surprise military attack is "among the greatest dangers a country can face—of the major wars in Europe, Asia, and the Middle East that have reshaped the international balance of power over the past several decades, most began with sudden attacks."[52] The challenge here with this kind of objective for target responders is that it is quite difficult to properly receive and process incoming information that both is sharply discontinuous from what has been received before and contradicts preconceived widely accepted notions about ongoing trends.

Strategic Manipulation Means

Strategic manipulation uses a wide variety of fluctuating techniques to achieve its ends. While requiring considerable adaptation by both initiators and targets of manipulation, the information explosion often "bestows the advantage" to initiators for "successfully using strategic deception to achieve surprise" and disadvantages those attempting to defend against such exposure.[53] Given that the offense has gained a decided (if not necessarily permanent) advantage, learning faster from past mistakes, the defense has typically felt rather over-whelmed.

Special circumstances seem to enhance strategic manipulation success. Following perceptual theories, strategic manipulation seems to work better when dealing with intangible intentions than with observable capabilities. If available, third-party channels can be highly attractive to deliver manipulated communication from initiators to targets. The best strategic manipulation initiators seem to be those at the bottom of the global hierarchy; the best strategic manipulation targets seem to be relatively secure and complacent countries (not those already highly skeptical about data accuracy). Indeed, great powers—given their expectations of dominance and universal rationality—tend to be both the most common and the most vulnerable targets, with democracies especially at risk; great power democracies often constitute not only the most favored foreign manipulation targets but also the most inept international manipulation initiators because of great internal pressures for foreign security policy openness and transparency. Within a vigilant global setting, strategic manipulation seems to work best if applied infrequently, because overuse and a well-established reputation for manipulation can lead to global distrust, unpredictability, and chaos, increasing target readiness to resist or thwart manipulative efforts and minimizing any meaningful long-term cooperation prospects within the international community.

Offensive Means for Manipulation Initiators

Strategic manipulation generally works through worsening the signal-to-noise ratio, lowering people's ability to perceive genuine new developments; facilitating distraction, misdirecting people's attention away from critical issues and causing analysts to waste resources on unimportant purposes; and catching targets vulnerable, nonresilient, and unable to predict global security perils. Offensive foreign manipulation often occurs through projecting an image of "daredevil" irrationality, reflecting "dramatic and brash decisions" that are "overreactive" to ongoing events; this pattern seems most likely to succeed when manipulation occurs within one's sphere of influence, at least partially to protect one's own citizens, and it can be quickly consolidated while surprise and a local power advantage persist.[54] Perhaps the most important strategic manipulation means is having foreign target background intelligence, especially on their psychological and organizational vulnerabilities,[55] because each potential target "has its own unique strengths, weaknesses, potential or actual biases, and vulnerability to manipulation."[56]

The specific means to pursue strategic ambiguity can entail sending messages through disconnected third parties, channels that the target perceives as somewhat unreliable, or government representatives saying that it is their "personal view" that a certain agreement might be acceptable to their home governments.[57] If initiators' goal is simply to confirm in a target's mind what it already expects, only a little ambiguous information is necessary; if the goal is to cause a target to change its mind, the mix of ambiguity and certainty must be more voluminous and sophisticated.[58] Thus knowing targets' preconceptions is vital so as to understand in advance how a target would interpret and react to ambiguous signals. If targeting enemy states, initiator language while containing vagaries about details may be bold and clear in its main thrust, especially when incorporating an ultimatum. Notably, state leaders often possess an advantage over private citizens in interpreting foreign government ambiguous signals; among states where political leaders understand tacit meanings of code words, the goal in employing ambiguity may be "to keep information from the general public, which lacks the keys to the code."[59]

Frequently war initiators intentionally use ambiguity to state objectives vaguely at the outset, since wars may require dynamic goal modification in response to changing battle conditions. For example, American general Maxwell Taylor remarked, "It is common practice for officials to define foreign policy goals in the broad generalities of peace, prosperity, cooperation, and good will—unimpeachable as ideals, but of little use in determining the specific objective we are likely to pursue and the time, place, and intensity of our efforts."[60] When confrontation objectives initially are articulated in vague and imprecise ways, they permit later strategic flexibility but are of lower value in guiding tactical combat.

Moving to the specific means to pursue strategic deception, it is "an umbrella concept" covering many different activities.[61] Deception may require concealment, exaggeration, equivocation, half-truths, and irony,[62] misdirecting targets' attention, causing targets to waste resources for unimportant purposes or catching them off-guard and unprepared.[63] Deception may work through reinforcing target preconceptions, altering them, or creating new beliefs where none had existed before.[64] Deceptive thrusts may cover capabilities, intentions, timing, location, magnitude of effort, and style of execution.[65] Deception may use "'leaks,' planted information, or decoys to create the impression that the truth is other than it actually is."[66] As to communication channels, sometimes intelligence agencies malevolently and covertly transmit "corrupt" information to foreign policy makers, and at other times "less clandestine methods of com-munication are used," including "media (television, radio, Internet) outlets, diplomatic interactions, academic exchanges and international travel and tour-ism."[67] Using such methods, "there is no single facet of the warning problem so unpredictable, and yet so potentially damaging in its effect, as deception."[68]

Certain circumstances promote strategic deception. Security issues involving intangible elements seem much more susceptible than those involving tangible visible elements. Cyberspace is highly conducive to deception, vital to launching cyberattacks, cloaking their identity, impeding precise attribution, and masking specific techniques used to penetrate targets.[69] Deception is better for short-run purposes; maintaining long-term secrecy is difficult in today's porous world be-cause of the proliferation of transnational watchdog groups (including outraged computer hackers) exposing deception and affecting security agendas. Global players widely viewed as untrustworthy, possibly because of past overuse of de-ception, are disadvantaged in deception initiation, because "the more one has a reputation for honesty, the easier it is to lie when one wants to."[70]

Finally, examining the specific means to pursue strategic surprise, opera-tionally it has many different facets, as colorfully noted by microeconomist Thomas Schelling:

> Surprise, when it happens to a government, is likely to be a complicated, diffuse, bureaucratic thing. It includes neglect of responsibility, but also responsibility so poorly defined or so ambiguously delegated that action gets lost. It includes gaps in intelligence, but also intelligence that, like a string of pearls too precious to wear, is too sensitive to give to those who need it. It includes the alarm that fails to work, but also the alarm that has gone off so often it has been discon-nected. It includes the unalert watchman, but also the one who knows he'll be chewed out by his superior if he gets higher authority out of bed. It includes the contingencies that occur to no one, but also those that everyone assumes some-body else is taking care of. It includes straightforward procrastination, but also decisions protracted by internal disagreement.[71]

What is unanticipated may be something other than simply the occurrence of a major event—"history is replete with instances in which the adversary was caught unawares by the timing, strength or location of the attack, even when the attack itself had been expected or considered a likelihood."[72]

Initiators' need to use strategic surprise varies inversely with their military advantage or the risks they would suffer from target retaliation,[73] making this approach most critical for weaker states. Thus if strategic surprise thwarts great powers, with surprise often neutralizing strong military forces' capability,[74] global "disappointment and resentment can occur."[75] However, despite the appeal of strategic surprise to lesser parties, especially when attacks involve short distances or do not require lots of advance preparation,[76] "genuine 'surprise attacks' are pretty rare today—at least at the strategic level—for the simple reason that mobilizing large military forces is a huge undertaking and hard to conceal."[77] Nonetheless, when surprise attacks do occur, they can be devastating—technological developments that have facilitated strategic surprise have also dramatically shortened the warning time for targets and determined war outcomes, thus threatening state survival.[78]

Defensive Means for Manipulation Targets

The defensive means to respond to foreign manipulation are usually low-key, attempting to avoid jumping to conclusions about the exact origins, intent, and impact of foreign influence attempts. To maintain mass public confidence in the target regime and to avoid any subsequent instability, political leaders often downplay seriously intrusive long-term disruptive foreign manipulation consequences. The underlying thrust behind such an understated defensive response is maintaining a target's sense of solid sovereignty, self-determination, and political independence. However, sometimes publicly minimizing the disruptive effects of foreign manipulation can backfire; fatalistic pessimism among target officials can promote state policy paralysis, undercutting political leaders' chances to react appropriately to early warning signals, and provoke anger about inattentiveness by citizens and the international community.

Defensive responses, specifically to strategic ambiguity, depend somewhat on the target. The typical national government response to initiator ambiguity is to try to find additional intelligence indicators to provide clues about how to properly interpret ambiguous messages received. The typical mass public response to initiator ambiguity, especially after private citizens have been victimized as a result, usually involves a call for more clarity and transparency from the national government about means and ends, as citizens expect accurate, complete details about how the national government is responding to threats.

Complicating defensive target responses is ambiguity—usually directly created and promoted by offensive initiators—in distinguishing between illegitimate foreign threats and legitimate international transactions. For example, in evaluating cross-national flows of people, goods, and services, such strategic ambiguity hinders target isolation of illicit activity; thanks to the manipulative efforts of transnational criminals, terrorists, and rogue states, "differentiating defensive firearms from assault weapons, legal refugees from illegal migrants, over-the-counter stimulants such as alcohol, tobacco, and coffee from illicit drugs, 'safe' substances from hazardous materials, mild sicknesses from lethal diseases, and free transmission of ideas from information disruptions involves considerable ambiguity in the eyes of governments and their citizenry and the international community as a whole."[79] In a parallel fashion, ambiguity surrounds target identification of state and nonstate manipulation perpetrators—various states label transnational terrorists and criminals differently, and these groups often do their best to enhance labeling ambiguity and to blend invisibly into the general population so as to avoid apprehension and prosecution. Such ambiguity can empower unruly disruptive forces, ones that sense sharp differences in onlookers' interpretations and subsequently attempt to exploit them to the fullest.[80]

Turning to defensive responses, specifically to strategic deception, targets could deceive attackers about attack outcomes, instead of attackers deceiving targets about attack origins.[81] An optimal target counterdeception mode is "to trick the deceiver into self-betrayal"[82] through creating "a unique environment that an attacker has never seen before," where it is confused, has to struggle to gain intelligence, is uncertain about the credibility of data obtained, and is "forced to lose time, wander into digital traps, and betray information" about its "identity and intentions."[83] In cyberspace, for example, targets can create this confusion through diversionary mechanisms such as *honeypots*, traps using a deceptively realistic computer environment to attract intruders, by which targets can monitor what attackers steal and obtain clues about their motives and identities.[84] Through these deceptive target mechanisms, a useless data-bank can be made to "look interesting as a way of persuading offenders to waste their time rummaging through it, show their cyber techniques to the defender, befuddle them with erroneous information, and perhaps get them to leave the system (falsely) satisfied."[85] Targets can then further confuse deception initiators about attack outcomes by using false damage reports, allowing attackers to see "only what is necessary to support the image that the defender wishes to project."[86]

Although universal vulnerability to deception exists, even among those whose officials are "sophisticated practitioners of the art themselves,"[87] some targets are more susceptible and more handicapped in responding than others.

Notably, "democracies are likely to be more vulnerable to deception than are dictatorships and closed societies and it is undeniably more difficult for open societies to practice it,"[88] for authoritarian states' "adeptness at deception and manipulation of information" is "facilitated by the closed and secretive nature of their societies."[89] The United States seems to be a frequent victim of foreign deception, typically as "a form of asymmetric warfare" used by opponents lacking the ability to win in direct military confrontations:[90] strategic deception "poses great danger to the United States," which is insufficiently attentive and skilled in dealing with this threat,[91] while transnational illicit nonstate groups such as criminals, rebels, and terrorists thrive on it because "all must seek cover to operate effectively,"[92] even though many "seldom have the resources, time or opportunity for strategic deception planning."[93]

In defensive responses specifically to strategic surprise, although the knee-jerk target reaction is to pour more resources into intelligence collection and analysis,[94] this is sometimes misguided because often surprise is not due to inadequate data: for example, a major recent empirical study of strategic surprise concludes that "the failure of senior officials to absorb and use information and analysis provided by professionals in the field can be more instrumental in creating strategic surprise than missing or faulty intelligence."[95] While "the common view is that surprise occurs because intelligence services fail to warn," usually "intelligence failures in collecting indicators were only secondary elements in the cause of surprise," for "leaders in the victim state were warned that the enemy was marshaling capabilities to strike, but they did not react to the warning in ways that hindsight demonstrates were necessary."[96]

As for the targets most susceptible to strategic surprise, major powers seem most attractive, because their preponderance of lethal firepower makes them difficult to overcome otherwise. Western states also appear highly vulnerable and slow to respond, tending to be flabbergasted and taken off-guard when discovering that their sophisticated warning systems fail or their enemies do not operate according to their concept of rationality.[97] The United States is an appealing target for foreign surprise, especially "when it focuses its diplomatic attention episodically on countries perceived to have immediate strategic value and engages narrowly and solely with elites."[98]

Strategic Manipulation Costs and Benefits

Although information overload generally facilitates strategic manipulation, even in this highly conducive context manipulation exhibits a mix of costs and benefits. For example, strategic ambiguity may be either "a source of anxiety, which decisionmakers seek to reduce," or "advantageous in permitting the maintenance of optimism and hope or preventing the premature closure of options."[99] So strategic manipulation's utility depends on identifying situations in which benefits outweigh costs.

Strategic Manipulation Costs

The following are general disadvantages of strategic manipulation:

- Its impact can degrade over time if repeated against the same target, used in the same way, or applied in too extreme a form.

- It can degrade the rights of citizens to witness clear and transparent policy making within democratic societies.

- It can degrade commitment credibility and meaningful alliances with desired global partners.

- It can degrade accountability for security policy decisions and increases the occurrence of plausible deniability.

- It can degrade the clean division of legitimate political authority between states and nonstate groups.

- It can degrade government officials' ability to prioritize security goals and threats and outsiders' ability to comprehend this prioritization.

- It can degrade the capacity to develop meaningful political, economic, and social acceptability norms.

If initiators use strategic manipulation improperly against foreign targets, it can lead to embarrassment, humiliation, and loss of global status, impeding trust and alliance relationships and occasionally even producing "relative diplomatic isolation."[100] If the cause of initiator manipulation failure is targets responding properly to foreign influence attempts, then these victims can reap substantial rewards in terms of resiliency and smoothly functioning foreign policy.

Regarding the specific costs of strategic ambiguity, it can be conflict-promoting, as it may foster confusion even among political leaders about communication meaning or what is actually transpiring, "prevent the signal from leading to an exchange in which both sides can move toward a compromise," and "create significant unfortunate and unintended misunderstandings among negotiators.[101] If agreements among contending parties are reached, ambiguity can reduce their value because of differing interpretations: popular but ambiguous terms such as *national security* "may not mean the same thing to different people," and indeed "while appearing to offer guidance and a basis for broad consensus they may be permitting everyone to label whatever policy he favors with an attractive and possibly deceptive name."[102] Given that adversaries generally exaggerate the clarity and appropriateness of their communication,[103] they often underestimate the ambiguity that targets may see and resulting interpretive differences, thus becoming quite mystified when misunderstandings and resentments ensue such as when an opponent's hostile responses are seen as overreactions. Ambiguity may also lower message credibility, because "ambiguous signals, 'unconfirmed' reports, and the so-called 'cry-wolf' syndrome tend to depress confidence in warning forecasts."[104] Ambiguity about the origins of security problems (with each side claiming that it is not really responsible) can lead to unproductive finger-pointing about whose burden it is to address troubles and to enhance "the unmanageable stress confronting national governments."[105] Indeed, ambiguity surrounding the division of authority between states and armed nonstate groups "may undercut the ability to hold either tightly accountable for their actions."[106] Most broadly, a source of civil society erosion, allowing exploitation of unprotected groups, can be systemic ambiguity about societal values—uncertainty about the rights of not only individuals and states but also subnational and transnational groups.[107]

Regarding the specific costs of strategic deception, when it is uncovered in politically sensitive areas, trust and the initiator-target relationship are degraded:

> Many operations needed to successfully counter low-intensity threats demand secrecy and sometimes, deception to mask politically sensitive training and deployment. . . . It is therefore tempting for government counter-insurgency agencies to engage in such effective but morally and politically ambiguous operations or "dirty tricks" hidden from the public. These "government secrets," if exposed, will certainly prove costly to the politicians governing the democratic country.[108]

Moreover, "nontransparency or deception in national security affairs—when done specifically in the context of government policies toward these violent non-state groups—could easily backfire and leave a state with little credibility in the eyes of its citizenry."[109] Manipulators can become particularly irked when they become victims of their own ploys,[110] and the negative fallout from uncovered political leaders' lies does not vanish quickly. China's deceptive actions vis-à-vis the outbreak of severe acute respiratory syndrome (SARS), an infectious respiratory illness that appeared in southern China in November 2002, spread to thirty-two countries, and killed eight hundred people before being contained in July 2003, illustrates these costs: "The way the Chinese regime attempted to cover up the situation and to disseminate falsehoods both domestically and internationally has seriously damaged its credibility."[111] If overused against one's own people, deception could even lead to regime collapse—several social uprisings against national governments have suggested that "regimes that only offer their people repression and delusion eventually collapse under the weight of their incompetence, corruption, and self-deception."[112] Generally, once international trust is eroded, deception can escalate, cross-unit negotiations can be hindered because of "the inability of actors to exchange information even when it is in the best interest of both sides,"[113] and, if deception is used too indiscriminately, "this could jeopardize credibility, alliances, and ultimately global survival."[114]

Regarding the specific costs of strategic surprise, its initiation can be highly risky, degrading predictable stability not only in relationships between initiators and targets but also within the international community as a whole. Relying too heavily over time on repeated strategic surprise as a means of achieving security objectives can be dangerous; for example, this could lead to hypervigilance on the part of strategic surprise targets, making it hard to execute such manipulation in the future, and the international community might become skeptical about what deficiencies in such initiators constantly prompted the application of strategic surprise. General costs surrounding strategic surprise include the possibilities of (1) leaks communicating intended secret plans to targets in advance, (2) backfire effects whereby trust in advance consultation with allies or domestic citizens is reduced, and (3) action-reaction cycles of impulsive surprise disruptions by multiple parties disadvantageous to everyone. Because of overconfidence, force initiators employing surprise are often likely to be frustrated when undesired outcomes ensue. Notably, "surprise attacks often fail disastrously because the side undertaking the initiative miscalculates in several ways," such as when "it leads the weaker side in the conflict to reach for goals that are truly beyond its grasp or to forget that when the effects of surprise dissipate, the dialectic of war returns with a vengeance."[115] If onlookers did not share an initiator's perceptions at the outset, they might view such an initiator as naïve or ill-informed for having misconstrued the ongoing predicament. For strategic targets, beyond the humiliating egg-on-the-face of being caught unaware of an impending threat, they are likely (because of wishful thinking) to be taken aback by the high scope of the damage, the unstable aftermath, and the inability to deal with dire consequences effectively. Citizen confidence would then decline precipitously in the ability of the surprised target national government to protect them from incoming threat.

Strategic Manipulation Benefits

In contrast, the following are the general advantages of strategic manipulation:

- It can, aided by information-overload-enhanced misperceptions, help promote constructive delusions that advance global peace, stability, and tolerance.

- It can enhance policy flexibility, leaving a wide range of policy options open and gaining the benefits from considering outside-the-box security policies.

- It can promote compromise agreements among contending parties with opposing interests, ones who otherwise would be highly unlikely to find common ground.

- It can facilitate saving lives and avoiding bloody confrontations by creating doubt in the minds of parties to the conflict about positive security outcomes.

- It can serve as an equalizer in lopsided international confrontations, giving weaker parties a chance against militarily superior opponents.

- It can stall or slow down the speed of decision making by targets as well as reduce targets' confidence in the appropriateness of policies undertaken.

- It can provide a safeguard against devastating revelations about intended security plans that must be kept secret.

If strategic manipulation is used properly against foreign targets, it can open avenues for future pursuit of national interests. If manipulation success results from targets mishandling incoming foreign influence, then target loss of societal control and internal and external respect seems likely.

Regarding the specific benefits of strategic ambiguity, leaving targets in doubt about what is transpiring allows initiators "to keep several paths open simultaneously and to initiate conversations without seriously endangering an image contradictory to the message sent and which the actor will want preserved if the other side's reaction is not favorable."[116] Especially in enemy relationships, ambiguity be useful in promoting coexistence when opposing interests exist, preventing preparation for a particular response, and allowing each side to spin outcomes as it pleases. Thus dispensing ambiguous information can give initiators much more flexibility than unambiguous information. Moreover, if one wishes to keep a target from reacting quickly to one's initiative, "ambiguity-producing tactics . . . are virtually always successful in producing procrastination—that is, in deferring appropriate decisions."[117] When weak parties engage in asymmetric conflict with major powers, ambiguity can blunt the value of superior military might, with the challenge being "sustaining the utility of military force in the absence of clarity, particularly when enemies discover that the use of protracted warfare in physically, ethically, legally, and politically complex environments offers protection."[118] When great powers confront each other, strategic ambiguity may be the best policy "to position oneself to avoid having to make invidious choices."[119] If used in initiating highly controversial actions, for which exact attribution of the source remains unlikely, ambiguity, inconsistency, and plausible deniability using third parties can be successfully exploited,[120] especially in cyberspace.[121] Most broadly, ambiguity can allow initiators to gain the advantages of appearing to be transparent without actually revealing significant details of their intended security plans, often desirable because of the need to ensure that sensitive information and covert threat response are kept secret.[122]

Regarding the specific benefits of strategic deception, perhaps its central value is that "it saves lives"[123] in a "relatively cheap" way.[124] Indeed, "even a primitive deception effort will, by threatening various alternatives, create enough uncertainty to distract the most wily opponent and force him either to disperse his effort or gamble on being right."[125] In coercive confrontations, deception can be valuable whether opponents are equal or unequal:

> Forgoing the use of deception in war undermines one's own strength. Therefore when all other elements of strength in war are roughly equal, deception will further amplify the available strength of a state—or allow it to use force more economically—by achieving a quicker victory at a lower cost with fewer casualties. In the case of unequal opponents, deception (and surprise) can help the weaker side compensate for its numerical or other inadequacies.[126]

In military affairs, deception is often "rewarded by greater achievements and success" than elsewhere,[127] such as when undertaking politically sensitive training and deployments or if needing to ensure survival when facing overwhelming odds.

Regarding the specific benefits of strategic surprise, in the diplomatic arena introducing surprise options can sometimes lead to quick compromise agreements among parties. In the military arena, surprise attack can yield enormous benefits and often lead to decisive victory:

> From a military point of view, the advantages to be derived from achieving strategic surprise are invaluable. A successful unanticipated attack will facilitate the destruction of a sizable portion of the enemy's forces at a lower cost to the attacker by throwing the inherently stronger defense psychologically off balance, and hence temporarily reducing his resistance. In compensating for the weaker position of the attacker, it will act as a force multiplier that may dramatically reverse the ratio of forces in the attacker's favor.[128]

Many advantages associated with first-resort use of force in surprise attacks:

> Waiting until other policies have failed may limit or forfeit the opportunity to use force effectively. The passage of time may mean the loss of surprise and the loss of initiative while giving the adversary opportunity to prepare military and politically for the battle to come. Also, waiting for diplomacy or, as is often the case, economic sanctions to work can allow people and other interests to suffer dearly.[129]

What is commonly seen as prudent patience in such situations can allow too much time for adversaries to anticipate and prepare for impending attacks, reducing the potential for success.

Assessing Strategic Manipulation's Cost-Benefit Balance

A review of the significant costs and benefits identified in this analysis does not lead to a sweeping conclusion about the overall utility of strategic manipulation under information overload. Instead, whether considering strategic ambiguity, strategic deception, or strategic surprise, the utility is decidedly circumstantial, varying by situation in a manner hinted at conceptually in this chapter and nailed down empirically in Chapter 4. However, undertaking strategic manipulation of foreign targets in today's world constitutes a high-risk/high-reward proposition, whereby possible constructive gains are huge but possible disastrous consequences are equally imposing. So extreme prudence is absolutely essential both in engaging in this kind of information and communication distortion and in responding to it.

3 Global Data Shock Case Studies

THE SELECTED TEN CASE STUDIES highlight distinctive security challenges for coping with global data shock, for both initiators using offensive manipulation and targets defending against manipulation under information overload. The cases are organized by theme—whether the primary form of manipulation exhibited by initiators is strategic ambiguity, deception, or surprise. Highlighting strategic ambiguity are the 2017 foreign security policy style of American president Donald Trump, the 2016 "Brexit" vote to leave the European Union, and the 2002–2003 nondiscovery of weapons of mass destruction in Iraq; highlighting strategic deception are the 2014 Russian annexation of Crimea, the 2011 Fukushima nuclear disaster, and the 2008 Russian invasion of Georgia; and highlighting strategic surprise are the 2007 Israeli attack on the Syrian al-Kibar nuclear plant, the 2005 Andijan massacre in Uzbekistan, the 2001 al-Qaeda terrorist attacks on the United States, and the 1990 Iraqi attack on Kuwait. Although by necessity every case rests on somewhat fragmentary evidence, this analysis makes every effort to unearth multiple sources of information about key insights in order to increase validity and reliability of findings and to draw balanced and dispassionate conclusions about the security impacts of information overload and strategic manipulation. While certainly not comprehensive, the cases were carefully chosen to be highly representative—in terms of type of geographical location, information overload, strategic manipulation, and defensive response—of the full set of global data shock patterns.

Each case is presented in parallel fashion, structured as follows:

- Description of ambiguity, deception, and/or surprise (including initiator, target, type of ambiguity/deception/surprise used, and dates of application)
- Role of information overload in triggering or changing ambiguity, deception, and/or surprise
- Rationale and purpose of the initiator for offensive ambiguity, deception, and/or surprise and by the target for its defensive response
- Policy effectiveness in the short run and the long run in the eyes of the initiator, target, and global community of ambiguity, deception, and/or surprise
- Perceived legitimacy in the eyes of the initiator, target, and global community of ambiguity, deception, and/or surprise
- Unintended short-term and long-term security consequences of ambiguity, deception, and/or surprise
- Lessons for initiators and targets for future management of ambiguity, deception, and/or surprise

This parallel case treatment is aimed at facilitating comparative cross-case analysis highlighting overarching patterns associated with strategic ambiguity, deception, and surprise under information overload, even though these manipulation techniques involve highly nuanced, subtle, and intangible forms of communication used in different ways by different societies.

Because the cases involve both manipulation success and manipulation failure, insights emerge about circumstances both when it seems best and worst to apply ambiguity, deception, and surprise under information overload. As would be expected, an inverse relationship exists between manipulation initiator and manipulation target success—what contributes to offensive success tends to increase defensive failure—and ironically it is often possible to glean as many if not more clues about needed security policy improvements from failures as from successes. The case findings will later suggest improvements in foreign manipulation management under information overload from the perspectives of both initiator offensive actions and target defensive responses. Critical to these improvements are the lessons identified at the end of each case emphasizing both (1) what was pivotally critical to achieve desired outcomes in cases in which they occurred and (2) what might have converted failure into success if appropriate steps had been taken in cases in which desired outcomes were thwarted.

Strategic Ambiguity Cases

2017 Foreign Security Policy Style of American President Donald Trump

President Donald Trump's foreign security policy has had one central over-arching thrust: "The Trump Administration is committed to a foreign policy focused on American interests and American national security."[1] Following his campaign rhetoric, he has aspired to "make America great again" by placing promoting United States security interests above all else. Trump has been specifically championing political, military, and economic "nationalism over international cooperation,"[2] believing that "America gets a raw deal from the liberal international order it helped to create and has led since World War II."[3] To accomplish this end, President Trump has revived President Ronald Reagan's idea of "peace through strength" as being at the center of foreign policy, believing that "the world will be more peaceful and more prosperous with a stronger and more respected America."[4] His top foreign policy priorities have been all security-oriented, including defeating ISIS and other terrorist groups and rebuilding American military capabilities.[5] Trump won the 2016 presidential election in part because "he sensed that the public wanted relief from the burdens of global leadership without losing the thrill of nationalist self-assertion."[6]

Description of Ambiguity, Deception, and/or Surprise   While Donald Trump campaigned for president of the United States in 2016, and after he was elected and assumed office in January 2017, his foreign security policy statements and actions consistently evidenced ambiguity in the eyes of both American citizens and foreign observers. This use of ambiguity, incorporating seemingly contradictory statements and actions, sudden policy reversals, and vague pronouncements open to widely differing interpretations, has appeared to be both intentional and strategic.

What makes this strategic ambiguity policy unusual and perhaps unprecedented among world leaders is how blatant and open is Trump's commitment to the unpredictability doctrine:

> Reversals and shifts are far from unprecedented. New administrations often adjust their policies to deal with the complex realities of international affairs or with changing tides in domestic politics. But few of these have openly sung the praises of unpredictability or contradicted themselves with such abandon as the Trump administration. The president and his supporters argue that having a reputation for being unpredictable will make others think twice before messing with the United States.[7]

Trump's form of strategic ambiguity is that you do not show your hand early in a negotiation, and then if others involved have no idea what you will do, and you display a willingness to do anything, you hold the advantage.[8] Although "ambiguity has always had a place in diplomacy," Trump adds to that "a free-style approach to international relations" reflecting "a disregard for norms and protocol, an impulsive nature and a tendency toward making contradictory statements."[9] While some observers have been baffled by this ambiguity, and others thought that it was simply a rhetorical device rather than a foreign policy strategy, it seems to be part of a preconceived plan.

Although not as front-and-center as his strategic ambiguity policy, President Trump also consciously used unpredictability to promote strategic surprise. In April 2017, this pattern was exemplified during an interaction between Trump and China's head of state:

> President Trump turned in his chair at Mar-a-Lago to get a better look at China's president, Xi Jinping—intent on detecting his first reaction to the news he had just dropped: American missiles were slamming into an airfield in northern Syria. It took a few moments, but Mr. Xi's eyes widened in surprise, and he asked his translator to repeat what was said, according to three people who spoke with Mr. Trump after that night two weeks ago. This was exactly the response he was hoping to elicit—surprise, uncertainty and a sense that the rational, predictable statecraft of President Barack Obama had given way to Mr. Trump's more assertive vision of American power . . . There are signs it has also made an impact on the Chinese, prodding them to finally use their leverage with their errant neighbor, North Korea.[10]

An unpredictability doctrine logically entails furthering one's agenda through strategic surprise.

Role of Information Overload in Triggering or Changing Ambiguity, Deception, and/or Surprise   Thanks primarily to the unprecedented role of the Internet and social media sites, during the 2016 American presidential campaign the American public was bombarded with more information about the two opposing candidates—Democrat Hillary Clinton and Republican Donald Trump—than ever before in American history. Much of the information received was of questionable credibility, difficult to pin down despite the presence of so-called fact-checking organizations, and so the result in many cases was mass public confusion and cross-cutting accusations of "fake news." The respected American television news program *60 Minutes* concluded that "in this last [2016] election, the nation was assaulted with imposters masquerading as reporters—they poisoned the conversation with lies on the left and on the right."[11]

After Trump assumed office in January 2016, the onslaught of sharply conflicting news reports and interpretations of his foreign security words and actions continued, enhanced by his own deliberate use of strategic ambiguity and surprise. Trump seemed to understand information overload and know how to operate in the emerging information milieu. Because assessments of Trump's foreign security policy performance often reached polar opposite conclusions, domestic and international confusion and misinterpretation intensified about his policies' coherence and credibility, including their content, interconnections, and goals.

One pervasive misinterpretation of Donald Trump's foreign security policy, primarily caused by its newness, contradictions, and vagueness under information overload—was its perceived inconsistency. He has been widely seen as someone who "is opportunistic and makes up his views as he goes along, but a careful reading of some of Trump's statements over three decades shows that he has a remarkably coherent and consistent worldview, one that is unlikely to change much" over time.[12] Indeed, "a closer read of Trumpian language" shows that "Trump has not necessarily flip-flopped on his policy positions" and that "it is primarily the inferences made about his remarks that are inconsistent with what he actually says."[13] Moreover, Trump "exploits the vagaries of communication," for his bombastic blasts are not meant to be taken literally.[14] Thus "beneath the bluster, the ego and the showmanship is the long-considered worldview of a man who has had problems with U.S. foreign policy for decades—Trump has thought long and hard about America's global role and he knows what he wants to do."[15] One can certainly strongly disagree with the appropriateness of Donald Trump's foreign security thrusts, but they are not as random or haphazard as they initially appear.

Rationale and Purpose of Ambiguity, Deception, and/or Surprise   Trump's use of strategic ambiguity and surprise has been designed to befuddle both domestic opponents and foreign adversaries. Before and after being elected, Trump "loves to give mixed signals—or, to use Trump's preferred terminology, he fetishizes 'unpredictability' in both domestic and foreign policy"—indeed, it has been "elevated to the status of political doctrine":[16]

> Trump promised throughout the campaign that he would be a more "unpredictable" commander in chief, accusing President Barack Obama of forecasting

his strategy to American enemies. "We are totally predictable," Trump said during a foreign policy address in April [2016]. "We tell everything. We're sending troops? We tell them. We're sending something else? We have a news conference. We have to be unpredictable, and we have to be unpredictable starting now."[17]

Unlike his predecessor President Barack Obama, who intentionally maximized predictability, prudence, and collaboration in foreign security policy because of his belief that this would promote global stability, peace, and cooperation, Donald Trump strongly contends that unpredictability keeps adversaries on their toes in terms of preparation for and reaction to American foreign security policy moves and thus dramatically increases the probability of success of these policies.

Policy Effectiveness of Ambiguity, Deception, and/or Surprise   In reaction to President Trump's intentional use of ambiguity and surprise in foreign security policy, outside assessments of his policy effectiveness are decidedly divided.

> As president, ambiguity is a high-risk doctrine, particularly on foreign policy matters. For every adversary Trump may try to keep on edge, he could engender confusion and anxiety from allies that rely on clarity and stability from the United States. Mixed messages from Washington could also provoke an unintended response from abroad with wide-ranging economic or security implications.[18]

Both supporting and opposing opinions about the value of Trump's ambiguity policy have been strongly emotionally charged, with optimism higher for short-term than for long-term benefits.

On the positive side, proponents of Trump's strategy of ambiguity and surprise have proclaimed it to be a resounding success:

> On foreign affairs . . . President Trump's strategic ambiguity has given his administration something impossible to quantify on a scorecard and more valuable than any one agreement or action. It has provided them with the diplomatic space to negotiate solutions to some of the world's most vexing challenges. . . . His overall strategic ambiguity and a lack of rigidity in foreign affairs presents an opportunity to solve problems which have confounded successive American presidential administrations. By calling out allies and showing an openness to adversaries, President Trump is keeping the world on its toes—just how he likes it.[19]

Even skeptics have sometimes admitted there is at least a future potential to take advantage of this kind of foreign security policy:

> Senior U.S. officials have been pondering how to take advantage of Trump's disregard for precedent . . . exploring whether this approach, reckless as it may appear, may open up possibilities for U.S. diplomacy. Several prominent foreign policy analysts have argued there's some benefit in creating ambiguity and uncertainty about what the U.S. might do abroad, especially after years of prudent and predictable Obama administration policy.[20]

Some of these skeptics have tried to identify particular limited circumstances under which Trump's strategic ambiguity and surprise approach might be most useful:

> There are situations where this might benefit American policymakers. If Washington wants to deter an adversary, but does not actually want to use force, then leaving the threat ambiguous reduces the political costs of backing down, stopping opponents at home from accusing you of chickening out of enforcing a supposed red line. If the goal is to keep an adversary from taking *any* provocative steps—even those short of what you consider worth using force or imposing sanctions over—then introducing some unpredictability about what would trigger a response might be a good idea.[21]

Ironically, despite the presence of ambiguity and unexpected changes in his foreign policy team's composition, Trump's supporters view Trump's foreign policy as displaying considerable resolve:

> While it is too early for definitive conclusions, President Trump's policies are better than skeptics predicted. Collectively, his talented foreign policy team is demonstrating attentiveness to moral, strategic and military threats and willingness to apply moral, strategic and military, as well as economic, pressure to deal with them. From U.N. Ambassador Nikki Haley's passionate statements against those who commit atrocities and WMD proliferators, to Secretary of Defense James Mattis' pronouncements on U.S. willingness to use force if necessary against the world's worst aggressors to National Security Adviser H.R. McMaster's and Secretary of State Rex Tillerson's calm resolve when talking with or about powers that would do us, our allies or civilians harm, to Trump's own harnessing of U.S. power and leadership to extract concessions from others, this administration's foreign policy is off to a promising start.[22]

One clear advantage of Trump's approach is the ability to act quickly and change course if necessary. For example, in early April 2017, President Bashar al-Assad of Syria dropped chemical weapons on rebel-held territory, and so President Trump responded right away by launching fifty-nine Tomahawk missiles into Syria:

> After six years of U.S. passivity, there are limits now, and America will enforce them. . . . Moreover, the very swiftness of the response carried a message to the wider world. Obama is gone. No more elaborate forensic investigations. No agonized presidential handwringing over the moral dilemmas of a fallen world. It took Obama 10 months to decide what to do in Afghanistan. It took Trump 63 hours to make Assad pay for his chemical-weapons duplicity. America demonstrated its capacity for swift, decisive action.[23]

So in this view "for all the roadblocks and headwinds President Trump has faced on the domestic front, there is little debate that he and his unconventional national security team have made a consequential impact on the course and conduct of foreign policy in his first 100 days in office."[24]

On the negative side, opponents of Trump's ambiguity and surprise policy claim it has been an abysmal failure, especially regarding American allies abroad. Foreign policy strategies based on ambiguity are traditionally maligned:

> The president...clearly prefers to be unpredictable. This can make sense as a tactic, but not as a strategy. Keeping foes off balance can be useful, but keeping friends and allies off balance is less so—especially friends and allies that have put their security in American hands for generations. The less steady they judge those hands to be, the more they may decide to look out for themselves, ignoring Washington's requests and considering side deals to protect their interests. Frequent policy reversals, even those that are welcome, come at a substantial cost to United States' credibility and to its reputation for reliability.[25]

One unconvinced observer warns that this approach "may create consternation among allies, even as it enhances deterrence of adversaries."[26] Another critic contends that "in a war, or a sports match, or even a reality show, there's something to be said for unpredictability—for keeping enemies (and audiences) on their toes—but in a friendship, this is deadly," for "no one wants a friend whose core diplomatic principle is flakiness."[27] Still others feel that "Trumpian unpredictability often undermines coercive diplomacy,"[28] sometimes criticizing specifically his Middle East policies.[29] Moreover, several analysts contend that the Trump administration has been overconfident about the value of strategic ambiguity:

> The president and his advisers, some of whom clearly like the fog [of ambiguity] . . . seem to imagine that it will help them govern just as it probably helped them win. They shouldn't be so confident. For legislators, too much fog is paralyzing. For voters, it's a recipe for nervous exhaustion. For allies, it's confusing; for enemies, it looks like an opportunity.[30]

Many vocal critics have objected because "ambiguity might encourage the adversary to probe your resolve and test the limits of your interests while making it more difficult to clearly signal that a particular move is a step too far and will credibly invite retaliation,"[31] with many strongly contending that foreign security policy ambiguity and surprise will eventually have to be abandoned.

Perceived Legitimacy of Ambiguity, Deception, and/or Surprise   The perceived legitimacy of President Trump's strategy of foreign security policy ambiguity and surprise has been widely questioned. Even within his administration, the frequency of unauthorized leaks to the press by July 2017 indicated some internal dissatisfaction.[32] The root of this internal and external legitimacy concern is the expectation of open transparency within a democracy, in which legislators, voters, and foreign allies all understand at least the general guidelines and directions of current and future foreign security policies. As one media commentator puts it, "In foreign policy, every American should demand predictability—and transparency—from the president," because "US national security depends on it."[33] Nonetheless, at least initially, this ambiguous foreign security policy temporarily increased Trump's popular approval—as of April 2017, "Mr. Trump's confrontational and improvisational approach to foreign affairs has lifted his mood, fortunes and poll numbers in recent days."[34] However, since that time, his popularity has plummeted.

Much of the mass media seem to have soured over time in reacting to Trump's foreign security policy ambiguity and surprise. Early on during the presidential election campaign, the media seemed fascinated by this unusual pattern: "On the campaign trail, the media loved Donald Trump's unpredictability—what would the wacky candidate do next?"[35] Yet shortly after Donald Trump became president, many of the same media analysts turned critical, claiming that "unpredictability isn't a strength," as "for a great power such as America, it's a recipe for instability, confusion, and self-inflicted harm to U.S.

interests abroad."[36] Although tolerance for hyperbole is usually higher for candidates during election campaigns than for heads of state in office, no specific explanation was forthcoming about this dramatic shift in the media's position on the legitimacy of Trump's foreign security policy.

Unintended Consequences of Ambiguity, Deception, and/or Surprise   As predicted by critics, a key unintended consequence of President Trump's use of ambiguity and surprise in foreign security policy has been a speedy loss of trust among several key allies. This has been evidenced in skepticism both that existing promises and commitments will be kept and that future arrangements can be formulated. Indeed, "anxious allies say that unpredictability might be better described as incoherence"—since assuming office, Trump "has held meetings with his counterparts from other countries," but "in some cases, those sessions have only heightened doubts that Trump has a clear sense of what direction he intends to take U.S. foreign policy."[37] One example of this confusion occurred during a White House meeting in March 2017 between President Trump and German chancellor Angela Merkel:

> Merkel tried to pin down Trump on one of the top concerns of U.S. trading partners: a proposed "border adjustment tax" to be imposed on imported goods." "Don't worry," Trump told Merkel, holding his thumb and forefinger close together. "It will only be a little bit." Trump's breezy answer—and Merkel's exasperation—has been the talk of diplomatic circles in Washington and Europe.[38]

Privately, many foreign ambassadors based in Washington, D.C., "complained—diplomatically, of course—that thin lines of communication have made it difficult for them to explain U.S. intentions to officials in their home capitals," "creating strain on traditionally solid alliances."[39] As one diplomat candidly stated, "It's quite distressing that the Americans are so unpredictable—unpredictability is the worst."[40] Although this loss of allies' trust is undoubtedly unintended, in many ways it is rather unsurprising, for one of Trump's key foreign security policy assumptions has been that "the United States is being taken advantage of by its allies," preferring "that the United States not have to defend other nations, but, if it does, he wants to get paid as much as possible for it."[41]

Lessons for Future Management of Ambiguity, Deception, and/or Surprise   One lesson is that an unexpected dramatic shift in foreign security policy style may not be universally counterproductive. Most observers have been too quick to dismiss ambiguity and surprise as useful foreign policy tools, ignoring the vast theoretical insights on strategic manipulation's effectiveness in both cooperative and conflictual relationships, and the empirical record of the vital roles of ambiguity and surprise in foreign diplomacy. Such criticism unfairly equates a doctrine of unpredictability with stupidity, ignorance, and inconsistency. Nonetheless, "Trump's seemingly cavalier and imprecise use of language does have a price—it helps turn a liberal order led by the United States into possibly a Hobbesian one where might makes right."[42]

A second lesson is that when there is substantial public dissatisfaction with foreign security policy success, evident at the end of the Obama administration, sometimes citizens simply desire a different approach, even one that could be just as problematic as the one it is replacing. Both Barack Obama and Donald Trump were elected based on platforms promising significant change. Most balanced analyses would conclude that the foreign security policies of neither George W. Bush nor Barack Obama came close to achieving their goals, so the mass public may deserve to witness the benefits and drawbacks of a completely contrasting strategy.

A third lesson is that an open press allowing public debate about a head of state's foreign security policy can only be beneficial. Such conflict-laden discourse reveals differing underlying ways to draw meaning from foreign policy statements and behavior concerning intentions and capabilities. This understanding can improve both state officials' and private citizens' ability to cope with information overload. Despite the confusion potential, an open press can sharpen our abilities to draw appropriate meaning about what is really transpiring in global security affairs.

2016 "Brexit" Vote to Leave the European Union

In 2016, concerns bubbled to the surface in the United Kingdom about continuing its membership in the European Union (EU). On June 23, 2016, in a state-sponsored referendum, the pro-Brexit position to leave the EU won with a narrow margin of victory—51.9 percent of the popular vote. Notably, this referendum "saw the highest number of votes ever recorded for a UK election."[43]

The vote for Britain to leave the European Union did not emerge out of nowhere. Indeed, "the process of UK withdrawal from the EU is, slowly and inexorably, bringing to the fore a fundamental clash of views on the purpose and achievements of the EU that has been simmering away . . . for decades."[44] In 2013, facing increasing euroskepticism within the Conservative Party, British prime minister David Cameron bowed to pressure, did what many analysts considered impossible to avoid,[45] and announced a future renegotiation of a settlement within the EU and an "in-out" public referendum, resulting in 2015 in the Conservative Party platform emphasizing "reform, renegotiation and referendum."[46] This was not the first time Britain had conducted a referendum on a crucial security issue; just a few years earlier the government had conducted a referendum on the issue of Scottish independence.

Description of Ambiguity, Deception, and/or Surprise   The Brexit case exhibited a mix of ambiguity, deception, and surprise. The strategic manipulation perpetrators orchestrating the "Vote Leave" campaign were not government officials but rather the private constituencies, mainly UKIP (the UK Independence Party), print media, and Internet-based social media. Months after the Brexit referendum, the British newspaper the *Telegraph* did a fact-checking study of the claims emanating from advocates both of remaining in and of leaving the EU during the heated campaign before the vote—the study found that of the eight major claims investigated, three had dead "wrong" accuracy (reflecting intentional deception), and five had "hard to say" accuracy (reflecting vagueness and ambiguity).[47] The deception was so severe that "misleading claims by the official campaigns in the EU referendum were widely seen as disrupting people's ability to make informed and deliberate choices"; consequently, the Electoral Reform Society recommended that "an official body—either the Electoral Commission or an appropriate alternative—should be empowered to intervene when overtly misleading information is disseminated by the official campaigns."[48] Moreover, the outcome of the Brexit referendum was clearly a surprise to both the British government and the international community: like Donald Trump's election as president of the United States later that year, professional pollsters badly missed the mark—"every single poll, even those within sampling error," overstated the proportion of the British popular vote for the status quo position of staying in the EU.[49]

Role of Information Overload in Triggering or Changing Ambiguity, Deception, and/or Surprise   Information overload about the decision whether or not to leave the European Union left British citizens at the mercy of those who wished to sway their opinion. Ironically, despite being inundated by manipulative material from both sides of the issue, thanks especially to the ambiguity "people felt consistently ill-informed."[50] In contrast to the heavily regulated broadcast media legally required to provide balanced news coverage, several influential and traditionally highly partisan British newspapers fought for leaving the EU with "ruthless determination."[51] In the prelude to the EU referendum, immigration became the central focus of discussion, with "the *Express*, *Mail* and *Telegraph* urging voters to 'take back control' of Britain's borders" and EU supporters bashed as "arguing for the status quo of self-interest and a lack of patriotism."[52] Media "coverage of immigration tripled over the course of the 10-week campaign" and was increasingly tied to economic burdens, with news headlines like "soaring cost of teaching migrant children" and "migrants cost Britain £17bn a year."[53] Overall, there is little doubt that "a largely EU-hostile UK press market has played a significant role in both feeding political negativity about the EU and having it reflected back in political discourse."[54] Notably, the media directly benefited in terms of revenues from mass public interest in this issue: "to the media, Brexit is the gift that keeps on giving—some four months after the vote, and after a spectacular reversal of fortune for David Cameron, George Osborne and others, Brexit remains front page news."[55]

Beyond traditional print media, Internet social media sites were a major instrument of manipulation via information overload; the "Vote Leave" organizers released "nearly a billion targeted digital adverts and spent approximately 98% of their money on digital campaigning."[56] Consensus exists that "strong social media campaigning was one of the most remarkable aspects of the EU referendum."[57] However, in the end "it is likely that mainstream media generated much of the news that was liked and shared on social networks—indirectly influencing people through sharing and via online discussion," and "mainstream media strongly influenced politicians and campaigners themselves, who devoted considerable time and energy to trying to shape the press agenda, or to attack opponents and defend their previous statements."[58]

This vote manipulation was full of emotional rhetoric, devoid of any quest for dispassionate objectivity. Thanks to this inflammatory rhetoric, "the campaign leading up to the vote to remain or leave the EU on 23 June 2016 was the UK's most divisive, hostile, negative and fear-provoking of the 21st century."[59] Particularly the immigration discussion was "a discourse of uncertainty and ever-growing anxiety, as well as xenophobia and hatred."[60] Consistent with the fear tactics, three nasty metaphors dominated news coverage of foreign migrants: "migrants as water ('floodgates', 'waves'), as animals or insects ('flocking', 'swarming') or as an invading force."[61] In the end, most analysts concluded that "the UK opted to leave the EU on 23 June 2016 after a campaign mired by scaremongering and the misuse of statistics."[62]

Rationale and Purpose of Ambiguity, Deception, and/or Surprise   The proponents of leaving the EU were motivated by a desire to address the growing British popular dissatisfaction with their plight, along with "discontent and of mistrust of the Establishment."[63] Specifically, between 2005 and 2016, in Britain "the perceived threats associated with EU membership intensified"; these included (1) the sovereignty threat, "caused by the integrationist and expansionist mentality of the EU"; (2) the economic threat, triggered by "the perceived failings of the EU as evidenced by the eurozone crisis, which acted as vindication in the minds of sceptics and justified their decade plus long argument about the folly of joining the single currency"; (3) the identity threat, stimulated "as a consequence of the rising salience of immigration in terms of voter concern" and "aligned to concerns about economic security as East European workers became readily blamed both for the unemployment and low wages of British workers, and for the over-burdening of already overstretched public services in an era of spending cuts"; and (4) the electoral threat, resulting from the rise of the UKIP "as a populist party representing 'the people' against the EU."[64] Brexit proponents consistently mined whatever pent-up anger the mass public was feeling, fueled by the slogan "Take Back Control," which "was powerful and vague enough to mean whatever you wanted it to."[65]

The British government was motivated by a desire to "stay the course" and maintain "business as usual." However, in the process state officials evidenced significant misperception in anticipating the surprise outcome of the Brexit referendum. In what some have called "the great miscalculation,"[66] Prime Minister David Cameron "underestimated the extent of Brexit sentiment that existed within the electorate as a whole," with the credibility of the "Remain" campaign undermined by significant Brexit advocacy within the British Conservative Party.[67] British government officials specifically assumed that "economy would trump immigration as the primary concern of voters," and this premise proved to be very wrong.[68] Those in positions of power paid insufficient attention to the rise throughout Europe of "political parties that aim to restore national autonomy, often by appealing to far-right, populist, and sometimes xenophobic sentiments."[69] Widely ignored was pervasive citizen concern that the fate of the United Kingdom was being too driven by outside forces, including foreign immigrants entering the country: "Many in the United Kingdom . . . pushed for a British exit from the EU, or Brexit, out of frustration with the number of British laws that have come from Brussels rather than Westminster."[70]

Furthermore, across time strategic ambiguity has characterized the British government's relationship to the European continent. The always eloquent Winston Churchill had said it best:

> The UK's connection with Europe has long been semi-detached. "We are with Europe, but not of it," Winston Churchill told the House of Commons in 1953 during a discussion of a proposed European defense community. "We are linked, but not combined. We are interested and associated, but not absorbed." . . . Churchill's statements employed the language of political ambiguity. Both proponents and opponents of Brexit appealed to Churchill's spirit during the referendum campaign. The most reasonable interpretation, somewhat surprisingly, was summed up by Boris Johnson, a leader of the Leave campaign and Britain's new foreign secretary: with regard to Europe, Churchill was pro having his cake—and pro eating it.[71]

This sense of political ambiguity, shared by many British citizens, complicated unqualified advocacy for remaining within the European Union, but at the same time sometimes "facilitates consensus on complex issues and helps keep leaders in power."[72]

Policy Effectiveness of Ambiguity, Deception, and/or Surprise   Ambiguity, deception, and surprise worked together to cause the Brexit referendum to turn in favor of those wishing to leave the European Union. The result was effective in stimulating the beginning of the EU departure process, even though the choice between "hard" (severing all ties between the UK and the EU) and "soft" (continued openness to foreign workers and smooth cross-national capital flows) exit options remained to be decided.[73] In the wake of the referendum outcome, "the transformation is palpable"—Britain is now prepared to undergo "the arduous negotiations" necessary to implement the decision to withdraw from the EU.[74] Both pro-Leave and pro-Remain forces would agree that in the end the Brexit referendum outcome altered the course of European politics.

The desirability of this outcome depends on one's position about the value of European Union membership. While some observers feel that the EU has fostered global cooperation and peace over the years and reduced interstate aggression, skeptics argue that the Brexit vote will decouple Britain from what has turned into a dysfunctional regional organization: "Crippled by the euro crisis and divisions over how to apportion refugees," the EU "no longer seems strong or united enough to address its domestic turmoil or the security threats on its borders," and as a result "national leaders across the continent are already turning inward, concluding that the best way to protect their countries is through more sovereignty, not less."[75] Although the European Union's erosion may not constitute a major global threat, because it is possible that "a Europe of newly assertive nation-states would be preferable to the disjointed, ineffectual, and unpopular EU of today,"[76] the Brexit vote's long-run security impact seems likely to be mixed at best and negative at worst for continental Europe.

Perceived Legitimacy of Ambiguity, Deception, and/or Surprise   Most legal scholars, along with Brexit supporters, have deemed the referendum to be quite legitimate and binding. For example, a post from the UK Constitutional Law Association argues that "it was quite proper for Parliament to put the question of whether the UK should remain a member of the EU to the British people for decision by way of a referendum."[77] Moreover, this position asserts that "political fairness and democratic principle require one to respect the outcome of the referendum even if one is persuaded that Brexit would be a very bad idea."[78] However, Brexit opponents found the whole idea of a referendum process as the basis for policy change to be illegitimate, calling the vote outcome "advisory,"[79] and have suggested ignoring or defying the vote. For example, one upset British citizen said, "The referendum is a nullity and represents yet another example of the catastrophic failure of governance that the Westminster system allows to happen all too often—so we lurch, ill-prepared, from one crisis to the next."[80]

Unintended Consequences of Ambiguity, Deception, and/or Surprise   In many ways, the most unexpected long-term consequences of the Brexit vote were more what did not happen rather than what did happen. The dire extremist postvote predictions of neither the "Leave" advocates nor the "Remain" advocates came to pass. For advocates of remaining in the EU, in the year following the referendum vote, housing prices did not drop precipitously, a British recession did not ensue, no emergency Brexit budget had to be created, and—perhaps most importantly—"the Bank of England, the OECD, the IMF, and the European Commission all revised their economic forecasts for the UK upwards, and said they were mistaken about the short-term impact of Brexit."[81] For advocates of leaving the EU, "prominent spokespeople in favour of leaving distanced themselves from some of the campaign's most striking promises, such as spending £350 million on the NHS [National Health Service], and from its claims about the falling rate of immigration to the UK after Brexit."[82]

Nonetheless, one area of genuine unexpected change is a likely shift in the nature of Great Britain's closest foreign ties. In the future, "Brexit could . . .

prompt the UK to reconnect with the Anglosphere and the wider world, opening up horizons closed off inside a regional bloc."[83]

Furthermore, the tone of British political discourse may have been irreparably transformed:

> Yet much of the acrimony, partiality and suspicions of dishonesty that characterised the campaign has remained. The consequences of the EU referendum campaign are still being played out, and will continue to be throughout the period that Britain negotiates its departure from the EU and beyond. The implications of a divisive, antagonistic and hyper-partisan campaign—by the campaigners themselves as much as by many national media outlets—is likely to shape British politics for the foreseeable future.[84]

Long after the Brexit vote, considerable background tension remains: "As the terms of the divorce hove into view, things could get very, very messy, if the initial skirmishes are an indication of things to come—expect history and identity to be running sores throughout the process."[85]

The Brexit vote was a key symptom of the weakness of the EU, and perhaps all regional organizations, in an age in which global bottom-up populist sentiments are on the rise:

> Europe currently finds itself in the throes of its worst political crisis since World War II. Across the continent, traditional political parties have lost their appeal as populist, Euroskeptical movements have attracted widespread support. Hopes for European unity seem to grow dimmer by the day. The euro crisis has exposed deep fault lines between Germany and debt-ridden southern European states, including Greece and Portugal. Germany and Italy have clashed on issues such as border controls and banking regulations. And on June 23, the United Kingdom became the first country in history to vote to leave the EU—a stunning blow to the bloc.[86]

Indeed, "the problem of defining Europe remains—does the EU's survival depend on the deeper and closer integration of a core group of countries?"[87] Growing skepticism and even disaffection among other EU members is thus a key unintended consequence of the Brexit vote.

Lessons for Future Management of Ambiguity, Deception, and/or Surprise   One Brexit lesson is how influential strategic ambiguity, deception, and surprise can be when emanating from a variety of nonstate sources targeting citizens focusing on extremely complex foreign security policy issues about which even government officials are somewhat ambivalent. Although normally state governments—possessing a wide variety of conventional political instruments for altering citizens' views—are the manipulation initiators, in this age of the Internet and social media, bottom-up digital communication can match and in many circumstances outclass top-down state efforts in actually getting through to and influencing target constituencies. This pattern seems especially likely when mass public economic dissatisfaction is really high and when the state itself is not unequivocally committed to pursuing the challenged foreign security policies.

A second lesson is that national government complacency about support for the status quo can be extremely dangerous in a world where extremist, nativist, xenophobic, and anti-establishment public sentiments appear to be flourishing. The fear of losing one's sustaining employment, one's cultural identity, and ultimately one's way of life can be palpable, and fast and furious globally induced changes are challenging all three so profoundly that the ensuing high-mass public distress invites outside manipulation. For defending against such manipulation, it is essential to develop deep comprehension and sensitivity regarding the nature of this discontent, and yet the British government—the target affected most by the strategic ambiguity, deception, and surprise initiated by advocates of leaving the EU—"didn't do enough to understand" the deep resentment and frustration of Brexit supporters.[88] That mistake ultimately proved fatal for the Cameron administration.

2002–2003 Nondiscovery of Weapons of Mass Destruction in Iraq

In the early twenty-first century, Americans were quite worried about the possibility of another violent attack from the tumultuous Middle East. Primary among these concerns was the fear that weapons of mass destruction (WMD) might be used against them: "After 9/11 and the anthrax attacks that followed in 2001, fear increased in the United States (US) and Western Europe concerning the use of non-conventional weapons such as chemical, biological, radiological and nuclear weapons (CBRN)."[89] Saddam Hussein, who had risen to power in Iraq in 1979, fought an eight-year war against Iran in the 1980s, invaded Kuwait in August 1990, oppressed his people and committed egregious human rights violations, and openly sought to gain hegemonic control of the Persian Gulf, was under particular scrutiny and suspicion. Thus not long after the 9/11 attacks on the United States, "the US and the United Kingdom (UK) together insisted that Iraq posed a threat to international security because of the alleged possession of WMD."[90]

As a result of this joint American and British concern, in November 2002 the United Nations Security Council passed Resolution 1441 giving Iraq a "final opportunity to comply with its disarmament obligations" and insisting that Saddam Hussein should allow UN weapons inspectors into Iraq to confirm the presence or absence of weapons of mass destruction.[91] During the UN weapons inspection team's time in Iraq, it could not find any weapons of mass destruction, but Hans Blix, head of the UN Monitoring, Verification and Inspection Commission, argued that his team still needed "months" to accurately assess the scope of Iraq's weapon arsenal.[92] Nonetheless, in March 2003, the U.S. government confidentially told the inspection team that it had to prematurely halt its still-incomplete work and leave the country right away.[93]

On March 20, 2003, American political leaders concluded that because they believed Iraq possessed weapons of mass destruction, an invasion was necessary,[94] and American and British troops entered Iraq to begin the second Gulf War. Despite claims that the invading coalition included forty-nine nations, the only other countries providing fighting forces were Australia and Poland (on land) and Denmark and Spain (at sea). The invasion concluded quickly, with Baghdad controlled on April 9, followed by the collapse of the Iraqi regime signified by the fall of Saddam Hussein's stronghold Tikrit on April 15. On May 1, 2003, President George W. Bush declared that major combat operations were over. Later, in December 2003, the United States captured Saddam Hussein himself, after which he was tried on war crimes charges, found guilty, and executed. As a result of the invasion and its violent aftermath, over 4,500 American soldiers died.[95]

Description of Ambiguity, Deception, and/or Surprise   The Iraqi government clearly and consistently initiated strategic ambiguity and deception to hide the true scope of its overall weapons capabilities and related intentions. The record of Iraqi ambiguity and deception toward the United Nations is particularly long and colorful. Following the 1991 Gulf War, the United Nations Special Commission was established to ensure Iraq's compliance with UN weapons policies regulating the production and use of WMD. In summer 1991, "Iraq unilaterally destroyed WMD equipment and documentation in an effort at concealment":[96] "an Iraqi document that fell into the inspectors' hands revealed that in April of 1991 a high-level Iraqi committee had ordered many of the country's WMD activities to be hidden from UN inspectors, even though compliance with the inspections was a condition for the lifting of economic sanctions imposed after the invasion of Kuwait."[97] The UN Special Commission concluded in 1998 that "Iraq had failed to provide a full account of its biological weapons program, thus leading many to believe the country still retained biological agents in several places."[98] By 2001, Saddam Hussein had refused several times to comply with UN demands and "resisted the UN Special Commission's attempts to destroy materials and facilities for the production of WMD."[99] At the same time, Iraq had demonstrated such a "maddening and deceptive attitude toward the UN inspection processes" that the commission concluded, "Iraq would never be forthcoming, and that if it was blocking access to the UN, then it must have something to hide":[100]

> Baghdad has become increasingly adept at hiding its weapons of mass destruction (WMD). Despite four years of intensive inspections, U.N. weapons inspectors did not discover Iraq's biological weapons program until they got a tip-off from Saddam's son-in-law, Hussein Kamel al-Majid, who defected in 1995. Even if inspectors destroyed all of Iraq's WMD facilities, as David Kay, the former chief nuclear weapons inspector on Iraq warns, "The weapons secrets are . . . well understood by a large stratum of Iraq's technical elite. . . . [Iraq] has now become more like post-Versailles Germany in its ability to maintain a weapons capability in the teeth of international inspections.[101]

Vividly displaying the use of contradictory signals to promote ambiguity under information overload, "Iraq told so many different stories over so many years to UN inspection teams that it became impossible for them to reconstruct an entirely consistent narrative; they simply could not keep the lies straight."[102] In the end, Saddam Hussein's "refusal to come clean about his stockpiles turned the U.N. inspection process into a deadly game of cat and mouse," in which "every passing day gives Saddam more opportunities to hide his arsenal."[103]

American government officials had their own reasons to suspect Iraqi weapons deception. Based on aerial and satellite intelligence, "US sources believed that Baghdad was storing a significant amount of chemical weapons in secret"[104] (later in 2004–2005 Iraqi chemical munitions manufactured before the 1991 Gulf War were discovered, but both the U.S. and the UN knew about this strategically insignificant stash).[105] An interim American investigation reported that "significant amounts of equipment and weapons-related activities" had been "concealed from UN inspectors."[106] Inspection teams also found "clandestine laboratories" and "live botulinum toxin," usable to make biological weapons, "at an Iraqi scientist's home."[107] In January 2003, John Negroponte, then American ambassador to the United Nations, stated that "there is still no evidence that Iraq has fundamentally changed its approach from one of deceit to a genuine attempt to be forthcoming in meeting [the UN Security Council's] demand that it disarm."[108] Skeptical Western observers believed that because "Saddam has proven to be the master of hiding things including himself,"[109] WMD could still have actually been in Iraq, just in locations nobody could find: American secretary of defense Donald Rumsfeld directly suggested that "if we cannot see or find the weapons of mass destruction, it means they are hidden even better than we thought."[110] A postinvasion report by the Iraq Survey Group (ISG), a fact-finding mission sent to the country to investigate the weapons claims, indicated that while "Iraq did retain prohibited WMD programs," these programs "were not so extensive, advanced, or threatening" as the American intelligence community indicated,[111] and that "Iraq planned to resume its [weapons] production as soon as the international sanctions against the country were lifted."[112]

Role of Information Overload in Triggering or Changing Ambiguity, Deception, and/or Surprise   Because of the American security focus on Iraq since the 1990 Kuwait invasion, and more broadly on the Middle East since the 9/11 al-Qaeda terrorist attacks, intelligence about Saddam Hussein and the spread of dangerous munitions in the region was relatively abundant prior to the 2003 Iraq War. Because this information overload (containing conflicting signals reflecting ambiguity) was combined with preconceived biases, both sides misinterpreted the other's intentions and capabilities: "The United States and Iraq developed images of each other through the privileged weighting of what were seen as especially dispositive pieces of information; subsequent information was interpreted in light of preexisting images; and the dynamic became such that images—and errors—become more rather than less entrenched over time."[113] As time passed, "the discrepancies between the U.S. and Iraqi views of reality, however, grew more, not less, divergent."[114] In the end, "it is truly remarkable that the United States and Iraq so grievously misread each other's capabilities and intentions after devoting more than a decade to taking the measure of the adversary."[115]

In the case of Iraq, Saddam Hussein "underestimated U.S. hostility" prior to the 2003 war:

He failed to appreciate the increased U.S. freedom of action after the collapse of the Soviet Union in the earlier war, and the decreased U.S. tolerance for the set of problems he represented after the terrorist attacks of September 11, 2001. Moreover, Saddam suffered from a general overestimation of the shared interests between Iraq and the United States, seeing the two countries as natural allies and himself as a useful bulwark against Iranian expansionism and radical Islamism more generally.[116]

In addition, Saddam Hussein "appears to have once again greatly overestimated Iraq's military capabilities and underestimated those of his enemies, repeating mistakes he made in 1980 (when Iraq invaded Iran) and in 1991" (when after invading Kuwait Iraq faced the United States and its allies in the Gulf War).[117] More specifically, he misestimated "the willingness of the Iraqi army to fight at the behest of the regime . . . and its ability to stand up to the U.S. military" and "underestimated the technological prowess of the U.S. armed forces."[118] Finally, Hussein "misperceived at the crucial moment the ability of 'friendly' states—especially China, France, and Russia—to restrain the United States from launching an attack."[119]

A few observers accused the American government of intentional deception in claiming that Iraq possessed weapons of mass destruction. In the alleged Bush conspiracy narrative, "the Bush administration deliberately sought ways to invade Iraq" and was "willfully conspiring to start a war in Iraq to further US interests"[120] through inventing false justifications. This narrative emphasized the heavy congressional and mass media criticism of the American intelligence community for "allowing the Bush administration to pressure it into exaggerating the Iraqi WMD program to justify the war":[121] the intelligence community's assessments of Iraqi WMD had "failed as a matter of rigorous analytic tradecraft to maintain multiple hypotheses against which to collect data," instead automatically assuming that Saddam Hussein "would be re-constituting WMD."[122] Supporting this narrative, a 2004 book even emerged claiming that "the Bush administration deliberately concealed the fact that it knew Saddam did not have an active WMD arsenal in February 2003, a month before the invasion of Iraq was launched," and that "the White House directed the CIA to fabricate a letter from an Iraqi intelligence official linking Hussein to the September 11 attacks" (afterward, "not a single official surfaced" to confirm any of these allegations).[123]

However, the reality was that even with its vast intelligence capabilities the American government was also a victim of misperception regarding Iraq. A thorough 2016 British report on the 2003 Iraq War's prelude set the record straight: the report "demolishes one of the certainties of some opponents of the Iraq War—that Bush lied his way into war"—and "shows the president believed Saddam possessed WMD, although he and his administration were irresponsible in their use of the intelligence that led them to that conclusion."[124] President George W. Bush and his advisors were vulnerable to human cognitive frailty, reflecting strong preexisting beliefs and cognitive consistency. They exhibited three specific processing distortions—"treating the worst-case scenario as fact," "glossing over ambiguities," and "fudging mistakes."[125] Bush's secretary of the treasury, Paul O'Neill, noted that "contingency planning" for an Iraq invasion had been considered as far back as President Bush's inauguration.[126] Moreover, in the months before the invasion "many Administration officials reacted strongly, negatively, and aggressively when presented with information or analysis that contradicted what they already believed about Iraq."[127] Given the American intelligence community's tendency to "overestimate the scope and progress of Iraq's WMD programs," the Bush administration "stretched those estimates to make a case not only for going to war but for doing so at once, rather than taking the time to build regional and international support for military action."[128] Notably, prior to the 2003 invasion, "*no one* doubted that Iraq had weapons of mass destruction"[129]—"no Western or Middle Eastern intelligence service is known to have dissented from Washington's assessment of Iraq's WMD capabilities and potential before the war."[130] What most likely happened is that a "combination of cynicism, naiveté, arrogance and ignorance"[131] led the Bush administration to engage in "cherry picking" intelligence that supported its position on Iraq to justify a decision to go to war that had already been made.[132] Thus the Bush administration was genuinely surprised when after extensive searching no weapons of mass destruction were found in Iraq.

More broadly, American policy makers "misread Saddam's perceptions of threat" by overestimating his focus on the United States:

> They found it difficult to understand that Saddam paid only intermittent attention to their policy toward him, and that he was concerned to a much greater degree with what he saw as the linked threat from Iran and Iraq's own Shiite majority. Many of his actions and signals on questions such as weapons of mass destruction, interpreted in the United States as evidence of dangerous malignity, were in fact directed at the Iranian/Iraqi Shiite threat and not intended for consumption by an American audience.[133]

Overall, "the interpretation of Iraqi actions by U.S. decisionmakers was driven by a particularly vivid type of theory-driven information processing—the formation of an enemy image of Iraq, organized around an inherent bad-faith interpretation of Iraqi actions"—which "quickly become resistant to disconfirmation."[134]

Rationale and Purpose of Ambiguity, Deception, and/or Surprise   Iraq's motivations for employing deception about the true scope of its weapons arsenal were to preserve the country's sovereign right to protect itself and keep its enemies uncertain about what they might face. After the war, Iraqi scientists confirmed that "the combination of U.S. bombing, U.N. inspections, disarmament efforts, unilateral destruction by Iraqi officials, and stiff U.N. sanctions had indeed eliminated Saddam's illicit weapons by the mid-'90s."[135] So the deceptive ruse that Iraq possessed a WMD threat was perpetuated for two reasons, both of which revolved around Saddam Hussein's fear that "admitting his WMD were gone would have shown a weakness that could have threatened his hold on power."[136] The first explanation revolves around deterring foreign attack:

> Iraq engaged in deliberate deception prior to and during the war. It is possible that Saddam Hussein hoped the deniable threat of CBW [chemical or biological warheads] use would deter a coalition attack on Iraq, without undermining subsequent efforts to have sanctions lifted. Thus, prior to the war, Iraq created stockpiles of chemical protective gear and equipment—perhaps to make such a threat credible (the stockpiles were later discovered by advancing coalition forces)—and may have fed foreign intelligence services false reports that it retained WMD.[137]

The second explanation focused on deterring internal instability within Iraq:

> Saddam has always evinced much greater concern for his internal position than for his external status. . . . He may have feared that if his internal adversaries realized that he no longer had the capability to use these weapons, they would try to move against him. In a similar vein, Saddam's standing among the Sunni elites who constituted his power base was linked to a great extent to his having made Iraq a regional power—which the elites saw as a product of Iraq's unconventional arsenal. Thus openly giving up his WMD could also have jeopardized his position with crucial supporters.[138]

Iraq thus used strategic deception to falsely project WMD capabilities to friends and foes alike.

The stated justification for the American invasion of Iraq included allegations that the Iraqi government produced weapons of mass destruction, had ties to the al-Qaeda terrorist group, and engaged in human rights violations, and as a result the primary initial goal was to rid the country of these arms and free the Iraqi people from the oppressive Saddam Hussein regime. General Tommy Franks, who commanded the American Iraq invasion force, detailed its purposes as follows:

> First, to end the regime of Saddam Hussein. Second, to identify, isolate and eliminate Iraq's weapons of mass destruction. Third, to search for, to capture and to drive out terrorists from that country. Fourth, to collect such intelligence as we can related to terrorist networks. Fifth, to collect such intelligence as we can related to the global network of illicit weapons of mass destruction. Sixth, to end sanctions and to immediately deliver humanitarian support to the displaced and to many needy Iraqi citizens. Seventh, to secure Iraq's oil resources, which belong to the Iraqi people. And last, to help the Iraqi people create conditions for a transition to a representative self-government.[139]

However, the most important motive undoubtedly revolved around the alleged WMD threat—in 2003 deputy secretary of defense Paul Wolfowitz told *Vanity Fair* that "for bureaucratic reasons we settled on one issue, weapons of mass destruction, because it was the one reason everyone could agree on."[140] White House press secretary Ari Fleischer specifically stated that the focus was on changing the Hussein regime and finding and destroying the chemical and biological weapons presumed to be in Iraq—"the definition of victory is those two factors."[141]

Policy Effectiveness of Ambiguity, Deception, and/or Surprise   Iraq successfully concealed from outsiders the true scope of its weapons program. Moreover, Iraq's claims of innocent blamelessness regarding WMD were globally vindicated, and the United States lost considerable global credibility as a result: "Because the WMD threat was publicly presented as the reason for invading Iraq, the large divergence between prewar descriptions of the threat and what had been discovered in the nine months since the war was a matter of some consequence."[142] Ironically, Iraq's most successful deception was not about hiding its weapons of mass destruction but rather about hiding its lack of weapons of mass destruction.

However, in the long run, the ensuing violent conflict served the interests of neither protagonist. For Saddam Hussein, the weapons deception ended up pushing the United States to invade Iraq, causing Hussein to lose the war, be removed from power, and subsequently be executed (on December 30, 2006), a rather steep price to pay for a small public relations victory. For George W. Bush, the thrill of the military victory against Iraq was more than outweighed by the agony of the unexpectedly widespread and destructive violent postwar chaos, which "set off a new wave of terrorism that has no end in sight."[143]

Perceived Legitimacy of Ambiguity, Deception, and/or Surprise   The United States and the West viewed Iraq's strategic deception and uncooperativeness with weapons inspection as frustrating and illegitimate. Despite Saddam Hussein's checkered past, there was an unfulfilled expectation of cooperation with UN weapons verification effects. The sense of illegitimacy fed suspicions of Iraqi WMD development.

The reaction to the American intelligence blunder was humiliation for the U.S. government abroad and outrage among its citizens. Facing intense political pressure and negative WMD blowback, President Bush formed the Commission on the Intelligence Capabilities of the United States Regarding Weapons of Mass Destruction, which on March 31, 2005, stated this:

> On the brink of war, and in front of the whole world, the United States government asserted that Saddam Hussein had reconstituted his nuclear weapons program, had biological weapons and mobile biological weapon production facilities, and had stockpiled and was producing chemical weapons. All of this was based on assessments of the U.S. Intelligence Community. And not one bit of it could be confirmed when the war was over . . . making this one of the most public—and most damaging—intelligence failures in recent American history.[144]

Senator Jay Rockefeller reflected citizens' ire when he said, "You just don't make decisions like we do and put our nation's youth at risk based upon something that appears not to have existed."[145]

Although nearly universally appalled (with the exception of the British government) by unsubstantiated American WMD claims, the international community was divided on the legitimacy of the Iraq invasion and its rationale. Prior to the American invasion of Iraq, North Atlantic Treaty Organization (NATO) members France, Germany, and Canada and non-NATO member Russia had opposed military intervention because they believed that "the level of risk to international security was too high," and they instead pressed for "disarmament through diplomacy."[146] After the war began, the United States received heavy criticism from Belgium, Russia, France, China, Germany, and the Arab League.

Unintended Consequences of Ambiguity, Deception, and/or Surprise   A key unintended consequence of the WMD debacle was increased distrust by American allies of American intelligence agencies claims and ultimately loss of credibility for American foreign security policy makers. Earlier, in March 2003, "when the United States and its coalition partners invaded Iraq, the American public and much of the rest of the world believed that after Saddam Hussein's regime sank, a vast flotsam of weapons of mass destruction would bob to the surface."[147] Afterward, when that did not occur, the United States had global egg on its face, and the CIA, whose intelligence served as a scapegoat for the invasion, was in the crosshairs:

> In order to defend the credibility of his agency, CIA Director George Tenet took the unusual step of issuing a statement last Friday [June 6, 2003] dismissing suggestions that the CIA politicized its intelligence. "Our role is to call it like we see it, to tell policymakers what we know, what we don't know, what we think and what we base it on. That's the code we live by."[148]

These credibility concerns were so significant that "reforms were put in place after the Iraq War to make it harder for suspect intelligence to bubble up to the top ranks without careful scrutiny (for instance, a new procedure required heads of intelligence agencies to vouch personally for the credibility of any of their own agency's sources that are used in a major estimate)."[149] Moreover, previous American cavalier accusations associated with the WMD claim had unnerved some allies:

> In his State of the Union speech on January 29, 2002, Bush proclaimed that Iran, North Korea and Iraq made up an "axis of evil" seeking WMD and involved in terrorism worldwide. While the phrase played well in America, U.S. allies and enemies perceived it as a proclamation that war was inevitable, and it frightened some of them into believing Bush was too much of a cowboy to be trusted on war plans.[150]

The enduring security result was difficulty for the United States in getting other countries to commit to actions supporting its foreign security policies.

Lessons for Future Management of Ambiguity, Deception, and/or Surprise   The primary lesson from this case is that a smaller, weaker state can often use strategic ambiguity and deception effectively, even against a superpower, especially when confirming a target's preexisting suspicions. Given the huge variety of Iraqi hiding places available to conceal weapons of mass destruction, the odds of a UN inspection team finding them expeditiously seemed minuscule; given that there were no significant WMD caches to find, Saddam Hussein's ability to keep the West guessing on this issue was impressive. Overall, the United States "spends more to run spy satellites and supersecret listening devices than the gross domestic products of many countries, yet it didn't have a clue as to what was really going on inside a sanctions-racked dictatorship it was about to attack."[151] The cleverest use of strategic ambiguity and deception is thus when the risk of detection is extremely low because of a tiny potential for the truth to be definitively discovered. At the same time, the American counterclaim that Iraq had weapons of mass destruction was quite high-risk because it demanded that unimpeachable evidence be found in a situation in which time and resources made this extremely unlikely.

Strategic Deception Cases

2014 Russian Annexation of Crimea

The coercive Russian takeover of Crimea, widely regarded as "a frontal challenge to the post–Cold War European regional order,"[152] has deep roots. In the nineteenth and twentieth centuries, the "Russification" of the Crimea was executed through massive resettlement of ethnic Russians and brutal deportation of Tatar, Greek, and Bulgarian minorities in the area.[153] After months of protest, including clashes from February 18 to February 20 involving about twenty-five thousand rioters with at least eighty-eight people killed in forty-eight hours, on February 22, 2014, violent Ukrainian protesters ousted Ukrainian president Viktor Yanukovych and seized control of presidential administration buildings. A few days later the parliament named Oleksandr Turchynov as interim president. Even after the Ukrainian parliament removed Yanukovych from power and he fled the country, he refused to resign, and politicians from the east and south (including Crimea) declared continued loyalty to him. Russia also refused to recognize the new interim government. On February 27, special police units (named *Berkut*), using armored personnel carriers, grenade launchers, assault rifles, and machine guns, seized key checkpoints, controlling all land traffic between Crimea and continental Ukraine; unmarked uniformed armed forces (later identified as Russian special forces) in the Crimean capital Simferopol seized the Crimean parliamentary building and the Council of Ministers building, replaced the Ukrainian flag with the Russian flag, and installed a pro-Russian politician, Sergey Aksyonov, as Crimea's prime minister.

With the Russian parliament's approval, Russian military units began moving into Crimea almost immediately after former president Yanukovych called on February 28 for Russian president Vladimir Putin to "restore order" in Ukraine. In response, on March 3, Ukraine mobilized its armed forces and reserves. On March 4, Vladimir Putin publicly announced his intention to annex Crimea,[154] and he ultimately installed what many see as a "puppet government" there.[155] After the Crimean parliament voted to join Russia on March 6, on March 16 a controversial referendum indicated that 97 percent of the voters chose to join Russia, but on March 27 a United Nations General Assembly resolution declared the Crimean referendum to be invalid. On June 7, Petro Poroshenko was sworn in as president of Ukraine, vowing to bring "peace to a united and free Ukraine."[156] In June, Russia finally announced the temporary withdrawal of regular combat troops from the border.[157] In the end, "Russia's occupation and annexation of the Crimean Peninsula in February and March have plunged Europe into one of its gravest crises since the end of the Cold War."[158]

Description of Ambiguity, Deception, and/or Surprise   Russian aggression in Crimea highlighted the West's strategic ambiguity policy toward the former Soviet bloc states surrounding Russia. For over two decades, "the corollary to giving Russia a 'voice without a veto' in Euro-Atlantic security affairs has been to offer Russia's neighbors a 'pledge without power'—to make promises which appear to convey binding security guarantees but without creating the mechanisms for their enforcement."[159] Specifically, NATO's vague promise at its Bucharest Summit in 2008 to offer Ukraine future membership in the alliance, with no specific date and no inclusion in the Membership Action Plan, was "misguided" because it was not preceded by "a careful examination of its medium and long-term security and political consequences and of the alliance's ability to bear their burden."[160] After the Russian annexation of Crimea, the West's strategic ambiguity persisted, as it imposed numerous sanctions on Russia—which seemed unlikely to achieve any significant changes—but "the true objectives of the sanctions have never been clearly stated."[161]

While the West was using strategic ambiguity, Russia—the strategic initiator in this case—employed strategic deception in its military action to conceal tactics and reduce international blowback. Much of the force that invaded Crimea consisted of masked "green men," "a hybrid between regular infantry and anti-terrorist police units, having a secret chain of command and bearing no insignia or visible rank on their combat fatigues."[162] Russia used covert military action, subversion, and subterfuge in pursuing its traditional military goal of "misleading the enemy with regard to the presence and dispositions of troops and military objectives."[163] The motive for this deception "was clearly designed to conceal the state identity of the invading force."[164] Russia also employed effectively subtle covert "non-military instruments of influence and diplomacy, which emphasized in particular a more or less plausible deniability in an effort to disable international responses and bolster domestic Russian support."[165] The wide variety of state-initiated nonmilitary instruments of information warfare included fake news stories, doctored photographs, and staged television clips.[166]

The Russian use of strategic deception was evident not only in its annexation of Crimea but also in its justification for the annexation, and a Russian disinformation campaign was an essential component in this ruse.[167] Specifically, "Russia drew on legal rhetoric to assist the process of 'deniable' intervention," attempting "to blur the legal and illegal, to create justificatory smokescreens, in part by exploiting some areas of uncertainty in international law, while making unfounded assertions of 'facts' (especially ostensible threats to Russians and Russian-speakers)."[168] Russia has a long legacy of using deception and disinformation in its foreign and domestic policies, including attempts (1) to link all internal problems to external causes, (2) to promote external enemy images, (3) to weaken critical thinking, (4) to facilitate societal consolidation in the face of military threat, (5) to create the view that the Russian head of state is the only one capable of resisting the military threat, (6) to prepare for inevitable wartime hardships, and (7) to propagate an image for the West of a united Russia ready for war.[169] Overall, "the Russian strategy, both at home and abroad, is to say there is no such thing as truth."[170]

Finally, the Russian takeover of Crimea generated strategic surprise in the West—the move "caught almost everyone off guard."[171] This annexation qualified "as a case of strategic surprise" because (1) "in both Crimea and Syria, Putin has sought to exploit surprise, moving fast to change facts on the ground before the West could stop him,"[172] and (2) "Western political leaders were overwhelmed by the development and tended to act in response to Russian actions rather than proactively" because of overconfidence, confusion, and signal "receptivity failure."[173] In retrospect, the Russian military action should not have been so unexpected to the West because it was "largely—if not primarily—conditioned by EU and NATO enlargement into the ex-Soviet space."[174] For example, in advance of the Third Eastern Partnership Summit in November 28–29,

2013, at which EU leaders met to witness the signing of a free trade agreement with Ukraine among other states, "Russia declared that it was willing to impose trade sanctions, energy supply interruptions, and security reprisals against states choosing to sign new agreements with the EU."[175] Although for many years Russian political leaders had "openly opposed Ukraine's integration into the economic, and hence also political, 'West' and in particular the possibility of its NATO membership," this well-known consistent Russian antagonistic stance had unfortunately been "regularly ignored by Western leaders who insisted on every European state's legal right to decide on its association with other states freely, including on membership either in EU or NATO."[176]

Role of Information Overload in Triggering or Changing Ambiguity, Deception, and/or Surprise   Information warfare has become a central part of Russian strategy. NATO reported "a significant increase in Russian propaganda and disinformation since Russia's illegal annexation of Crimea in 2014," with Russia's response to this accusation being that it simply presents "an alternative viewpoint that is ignored by the mainstream Western media."[177] The advent of the Internet, social media, and the digital revolution has made Russian political manipulation much easier than in the past. The details are truly harrowing:

> Where once the KGB [Russian intelligence agency] would have spent months, or years, carefully planting well-made forgeries through covert agents in the west, the new *dezinformatsiya* is cheap, crass and quick: created in a few seconds and thrown online. The aim seems less to establish alternative truths than to spread confusion about the status of truth. In a similar vein, the aim of the professional pro-Putin online trolls who haunt website comment sections is to make any constructive conversation impossible.... Taken together, all these efforts constitute a kind of linguistic sabotage of the infrastructure of reason: if the very possibility of rational argument is submerged in a fog of uncertainty, there are no grounds for debate—and the public can be expected to decide that there is no point in trying to decide the winner, or even bothering to listen.[178]

The Kremlin uses "this mix of political technology, fluid ideology, and corruption not only domestically but also in managing foreign relations."[179] The West seems an ideal target for such manipulation, for there is "increased uncertainty about collective identities and a perceived loss of control," leading to openness to conspiracy theories "especially rife among supporters of rightwing nationalist and populist parties."[180] Indeed, "the underlying goal of the Kremlin's propaganda is to engender cynicism in the population"—"when people stop trusting any institutions or having any firmly held values, they can easily accept a conspiratorial vision of the world."[181]

Vladimir Putin has consistently excelled at using the media "to consolidate power inside Russia and, increasingly, to wage an information war against the West."[182] For example, "Putin succeeded in augmenting decision uncertainty for the West" through sending "inconsistent signals by presenting himself at the international negotiation table as being engaged in finding a peaceful solution to the current conflict while simultaneously increasing military presence in the region and supporting the separatists in Ukraine."[183] What made the Russian approach to information warfare distinctive was its "new stress on the 'psychosphere' as the theatre of conflict"[184] and its "language of emotions and judgments, and not of facts."[185] Partially because of the flood of conflicting information, Putin could provide clear and satisfying distorted meaning that deflected attention away from his regime's shortcomings: "Today, Mr. Putin presents Russia's actions as responsive, not aggressive—every time Russia attacks a former Soviet republic, the confrontation is portrayed as a proxy war started by America against Russia."[186] Although Western observers found this narrative outlandish, many Russian citizens appeared to swallow it completely.

Rationale and Purpose of Ambiguity, Deception, and/or Surprise   Russia had several motivations to annex Crimea: (1) emphasizing geopolitics and strategic goals in its competition with the West, Russia wanted to assert its primacy in its region over its former satellite states; (2) emphasizing Russian identity and status, Russia wanted to affirm its importance as a protector of ethnic Russian communities; and (3) emphasizing domestic regime consolidation, Russia wanted to prevent the internal spread of populist protests pushing political alternatives as part of "a continuing effort to quash the opposition to Putin's centralized rule."[187] In the wake of the largest mass protests since the 1990s in the winter of 2011–2012, the Russian military deployment in Ukraine reacted to "Putin's own fears of growing western influence in eastern Europe."[188] Over time, "through a strategic combination of propaganda and geopolitical aggression, Putin's government has promoted a narrative meant to bolster patriotism, and Russian xenophobia and paranoia along with it."[189]

Russian plans to annex Crimea developed at least two decades before this crisis, with NATO's 2008 announcement to offer Ukraine future membership sealing the deal.[190] After the ousting of President Yanukovych on February 22,

2014, "the temporary power vacuum, the state takeover by groups supported and some financed by the West, and the general confusion offered an ideal opportunity for the Kremlin to carry out the latest version of its contingency plans for annexing Crimea."[191] Russian perceptions of the new Ukrainian leadership "as especially hostile to Russia and its political system seemed to trigger strategic worst-case thinking," leading to "a rapid decision to implement a plan for the occupation and eventual annexation of Crimea."[192] Once it began, the coercive Crimean annexation was "obviously well-prepared, rehearsed in advance and professionally executed."[193] Constant since the Soviet Union's fall is "Russia's paternalistic view of its post-Soviet neighbors—Russia continues to regard them as making up a Russian sphere of influence" where it has "privileged interests."[194] This condescending attitude reinforced special Russian fears of losing the "strategically important" Sevastopol naval base.[195]

The rationale for the West's strategic ambiguity policy toward Russia's surrounding satellite countries at the time of the Crimea annexation is complex:

> Strategic ambiguity was designed to square three very different circles. The first was how to reassure countries newly freed from the Soviet yoke that, as Russian power resurged, they would not find themselves under threat from Moscow. The unhappy experience of the three Baltic States—which enjoyed twenty years of independence between the two World Wars only to be incorporated into the Soviet Union—drove efforts to seek binding security guarantees from the Western powers. The second was how to avoid complicating U.S. (and European) relations with a Russia that might, under the right circumstances, become a true partner to and even member of the Euro-Atlantic world. The final and perhaps most decisive consideration was to avoid taking on burdensome new obligations or political costs.[196]

None of these considerations was made explicit. Moreover, "the Ukrainian nationalists' hopes that the West's sanctions against Russia will resolve the problem" were "utterly unrealistic."[197]

Policy Effectiveness of Ambiguity, Deception, and/or Surprise   The Russian annexation of Crimea must be deemed at least a short-run military success: "The case of Crimea stands apart from Georgia in 2008 by the effectiveness of the combined pressure of all the tools employed," as "the efficacy with which these achieved Russia's political goal was unprecedented, partially because observers had problems distinguishing between peace and conflict."[198] Within just days, and "with almost no bloodshed, Ukrainian troops on the peninsula had been contained or won over, Crimean politicians had been induced to dissolve the parliament and replace the prime minister with a member of Crimea's Russian Unity party, and the region had been 'reunited' with Russia."[199] Crimea remains under Russia's thumb through the present day. However, judged by long-run political consequences, the annexation's effectiveness is much murkier, as "its political aspects at times revealed an almost farcical lack of preparation."[200] Moreover, "by annexing Crimea and threatening deeper military intervention in eastern Ukraine, Russia will only bolster Ukrainian nationalism and push Kiev closer to Europe, while causing other post-Soviet states to question the wisdom of a close alignment with Moscow."[201]

For many analysts, the Russian annexation of Crimea represented the failure of the West's strategic ambiguity policy: it is "the final nail in the coffin of a Western strategy of 'strategic ambiguity' with regards to the states of post-Soviet Eurasia—an approach that had already been seriously compromised in the wake of the 2008 clash between Russia and Georgia."[202] With NATO attempting to make its policy toward its eastern boundaries "fuzzy," Russia called its bluff,[203] causing the lack of firm united commitment to act to be revealed for all to see. Moreover, this Western policy failure clearly resulted from misperception of Russian intentions: "In the years leading up to the Ukraine crisis, Western governments fundamentally misread Russia—they mistook Moscow's failure to block the post–Cold War order as support for it, assumed that Russia's integration into the world economy would invest the country in the status quo, and failed to see that although few Russians longed for a return to Soviet communism, most were nostalgic for superpower status."[204]

Perceived Legitimacy of Ambiguity, Deception, and/or Surprise   The international community reacted quickly and negatively to Russia's occupation and annexation of Crimea. On March 27, 2014 the United Nations condemned the Crimean annexation, and on April 1, 2014, NATO did the same thing and declared it to be "illegal and illegitimate."[205] Many outside analysts viewed stopping this kind of behavior as essential to future regional stability: "To reinforce the principles which underpin European security . . . it is essential to assess and refute unjustified Russian legal claims which seek to deflect attention from Moscow's use of force and seizure of territory" because "otherwise Russia may be ready to stake out a wider legal/normative challenge to western states beyond the clashes in spring and summer 2014."[206] Many Western countries imposed economic sanctions against Russia. However, these sanctions may have backfired: "US and European sanctions hurt the Russian economy and raised food prices," but "these misfortunes—which have hit the middle and lower classes much harder than they have hit oligarchs—encourage the Russian people to see the west as an implacable enemy."[207] In any case, the West has misjudged its "ability to coerce Putin through sanctions and diplomatic isolation—sanctions have not altered Russia's behavior in eastern Ukraine, and few experts think that any financial penalty could be great enough to convince Moscow to hand back Crimea."[208]

Despite the global outrage at its military action, Russia adamantly defended its legitimacy:

> President Vladimir Putin and official Russian propaganda used the right of the Crimean people to self-determination in the form of secession as the chief argument to justify and legitimise the annexation. Russia's much stronger historic claim to Crimea was also stated. Russia conquered Crimea and *de facto* possessed it much longer than Ukraine (for around 168 years vs. 60 years). In his Presidential address to the Federal Assembly on December 4, 2014, Vladimir Putin stressed the strategic importance of the peninsula also as "the spiritual source" of the Russian nation and state, citing the fact that Grand Prince Vladimir was baptized in Herson. According to Putin's claim, Crimea has had "invaluable civilisational and even sacral importance for Russia, like the Temple Mount in Jerusalem for the followers of Islam and Judaism."[209]

Russia "cloaked its actions in legal language, as other major states have done in the past, with the aim of fostering a reputation as a lawful actor."[210] Notably, in the aftermath of the Russian annexation of Crimea, the Russian people seemed to largely accept their government's story: "Putin's campaign to make Russia great again has been a ringing success, at least for domestic approval ratings."[211] In the end, "the Kremlin's current strategy for keeping control—manipulating all facets of the political process, adopting whatever ideological stance is expedient for a given situation, and buying loyalty with money and favors—has created a cynical citizenry, shaped by propaganda and conspiracy theories, that is bereft of hope" and "leaves Russia's international adversaries and allies alike uncertain of what to expect."[212]

Unintended Consequences of Ambiguity, Deception, and/or Surprise   Several unintended negative side effects of the Russian annexation of Crimea have occurred in its aftermath. Within Crimea, internal troubles have escalated:

The abrupt separation from Ukraine has created a number of serious problems for Crimea due to its high dependency on Ukraine for water, electricity, rail and road connections to the mainland among other things. Crimea lost about two thirds of its budget revenue, which used to come from the Ukrainian central budget. The separation also entailed the loss of about 70 percent of all tourists from Ukraine while Russian and other tourists might not substitute the loss soon and to a high degree. Many Crimean companies became deprived of their access to European markets. The Crimean authorities have faced huge problems with issuing new citizenship, ownership and property documents, as Ukraine blocked access to the central registries and the Crimean government did not have its own records. The replacement of the Ukrainian *hrivna* as currency with the Russian *ruble* has created additional disturbances in the economy and for the Crimean population. An important motivation for secession—the expected rise in the standard of living—has not, so far, materialized. . . . Members of the Ukrainian and Tatar minorities have also experienced pressures, dismissals and threats from Russian nationalists.[213]

Because those in Crimea supporting the Russian takeover did not fully anticipate all of these negative consequences, in the long run public dissatisfaction within Crimea and between Crimea and Ukraine may well escalate.

As to international relations, the ineffective Western response to Russian annexation of Crimea "has worsened the general political climate in the Euro-Atlantic area."[214] The crisis highlighted transatlantic community fragmentation: Western states initially agreed that 'the territorial integrity of Ukraine' was at stake on Crimea," but "the governments slightly differed from what extent international, European, or even national values were threatened by the annexation."[215] In particular, "while the EU, Britain, and Germany emphasised the threat to peace and security 'in the heart of Europe,' France and the US underlined international law and the stability of the international order."[216] Russian aggression "exposed deep divisions" between Europe and the United States, in part because of European states' fears that their security is no longer central to American strategy.[217] Furthermore, despite NATO's recent "ramping up its own military presence in eastern Europe,"[218] the weak Western reaction to Russian action significantly increased "the sense of insecurity in states bordering on the Russian Federation, particularly those having within their borders Russian minorities."[219] More broadly, growing regional anarchy is a real possibility because of Russian territorial aggression occurring with impunity—"Putin has evoked the scenario of a world where states 'live without any rules at all' and where there are heightened risks from internal instability in states, 'especially when we talk about nations located at the intersection of major states geopolitical interests.'"[220]

Lessons for Future Management of Ambiguity, Deception, and/or Surprise One lesson from Russia's annexation of Crimea is that it is unwise to use strategic ambiguity against a strong, determined, and ruthless target that has a proclivity to call one's bluff and respond with brute force. When NATO selected Russia as its strategic ambiguity target regarding the eastern borders of the alliance, it was a decidedly poor choice because Russia is exactly the type of country that would respond to this vagueness with action that tested the limits of NATO's actual willingness to commit forces to prevent Russian expansion.

A second lesson is that deception and surprise may work better when operating within one's own recognized sphere of influence. The West found itself lacking the tools and the credibility to convince nearby Russian propaganda targets that what they were hearing was highly suspect. Having ethnic Russians previously resettled in Crimea aided in this persuasion resistance.

A third lesson is that the digital revolution and increasing social media prominence can facilitate unscrupulous initiators applying strategic manipulation at home and abroad. For Russia these tools helped immensely in its disinformation campaign during the Ukraine crisis. The Russian state deception experience and expertise overwhelmed any counterefforts.

A fourth lesson is that responding to an adversary's use of deception and surprise by undertaking negative economic and military sanctions is never a great idea if the adversary is capable of convincing its people that any subsequent citizen suffering is the fault of these foreign-imposed sanctions and if the adversary's political leaders can remain untouched by the sanction's direct consequences. The only ones truly suffering from Western sanctions were the Russian people themselves, and that served only to increase their anti-Western hostility, not to pressure the Russian government to withdraw from Crimea.

2011 Fukushima Nuclear Disaster

With visible evidence of the devastating effects of nuclear radiation through accidents like the 1979 Three Mile Island nuclear disaster in Pennsylvania, and the 1986 Chernobyl nuclear disaster in Ukraine, the world has come to fear peacetime nuclear accidents almost as much as wartime nuclear bomb attacks.

However, even with heightened government-sponsored safety precautions, nobody expected the "triple catastrophe"[221]—earthquake, tsunami, and nuclear meltdown—that occurred as Japan's Fukushima disaster.

On March 11, 2011 at 2:46 p.m., the Great East Japan Earthquake—the largest in Japan's recorded history[222] with a magnitude of 9.0—hit seventy miles off Japan's northeastern coast, triggering a tsunami and a nuclear accident at the Fukushima Daiichi Nuclear Power Plant, owned and operated by the Tokyo Electric Power Company (TEPCO). The seismic tremors knocked out electricity, gas and water supplies, and telecommunication. The tsunami, which towered up to 12 meters high and reached points up to 40 meters (130 feet) above sea level,[223] arrived around forty minutes later, breaching the nuclear plant's seawalls and disabling emergency backup power generators and seawater cooling pumps, leaving only battery power to run cooling systems, which ran out in a few short hours.[224] The tsunami caused seawater to inundate the area, destroying or washing away buildings, vehicles, heavy machinery, and oil tanks, as well as overwhelming supposedly safe evacuation zones and eradicating entire coastal communities.[225] A triple nuclear meltdown occurred over three days after the earthquake and tsunami hit, as the earthquake and tsunami had seriously damaged the six nuclear reactors at the Fukushima plant: "When the earthquake hit, the plant lost power and water used to cool the reactors stopped circulating; back-up diesel generators located in the basements failed when the tsunami flooded the buildings; without power to cool the water which controls the temperature of the fuel rods, the fuel melted, turning into molten lava and breaching the steel containment vessels at the reactors' floor."[226] Afterward, at reactors 1, 3 and 4, hydrogen built up as the water used to cool the fuel rods became very hot, producing four hydrogen explosions that blew the roof off the nuclear reactors and released massive amounts of toxic radiation into the atmosphere.

What happened in March 2011 was unprecedented in human history, and as a result Japan found itself "at the center of its most dramatic crisis since World War II."[227] In retrospect, "the nuclear disaster caused by the earthquake and tsunami has been rated by the International Atomic Energy Agency as equal in severity to the 1986 accident at Chernobyl, the worst nuclear disaster on record."[228] Notably, "Japan's plants are all located in coastal areas, making them especially vulnerable to both quakes and tsunamis."[229] The scope of the devastation was not just "the thousands of people who were killed, and the people who were made sick by radiation sickness and will die within decades," but also the tragedy of a "beautiful region of the country that's been decimated for many hundreds of years."[230] In 2012, Dr. Michio Kaku, professor of physics at New York City University, investigated the condition of the reactors, confirming that reactor 2 had liquefied, and he declared, "We have never seen this before in the history of nuclear power—a 100% liquification of a uranium nuclear core."[231] In the wake of the disaster, hundreds of thousands of people ended up homeless and lacking clean water, electricity, or heat.[232] Over twenty-eight thousand people ended up being killed or went missing, and "more than 196,000 homes and other buildings have been totally or partially damaged."[233] The economic cost of damages, including the wreckage of buildings, lifeline utilities, and social infrastructure, is estimated to be approximately 16.9 trillion yen,[234] with physical damage alone costing Japan from $195 billion to as much as $305 billion (roughly equivalent to the GDP of Greece and twice that of New Zealand).[235] Airborne radioactive particles of plutonium reached not only Tokyo but as far away as Seattle, Washington, in the United States and Lithuania in central Europe.

Description of Ambiguity, Deception, and/or Surprise   In the wake of the Fukushima disaster, the Japanese government used strategic deception toward its own citizens and toward the international community, largely through omission of absolutely critical details. Although "the Japanese government knew at the time and immediately told the U.S. military exactly what had happened—they did not tell the Japanese people for three months,"[236] with "information on radiation risks only gradually and patchily released."[237] While Japan's prime minister on March 15 told TEPCO not to withdraw its nuclear plant workers so as to avoid a nuclear catastrophe, and Japanese government officials told Japanese citizens living within the 20-to-30-kilometer-ring to stay indoors, the U.S. government was urging all American citizens in the area to flee at least 80 kilometers away, and the French government was urging all its citizens as far away as Tokyo (250 kilometers away) to evacuate.[238] Moreover, the Japanese government promoted a tragic but laughable propaganda lecture by Dr. Shunichi Yamashita—the government's Fukushima Radiation Health Risk Advisor—on March 21, 2011, in Fukushima City; included in his talk were senseless comments such as "radiation doesn't affect people who are smiling," "happy drinkers are less affected by radiation," and "adults over 20 years have very little sensitivity to radiation."[239] In the words of a Fukushima resident who fled the area, initially "key information was being concealed completely," as "the Japanese government didn't ask for local residents to evacuate, but instead paid them to dig, mow, cut down trees and clean buildings with water" ("many people died because they did 'the cleaning'"), and "Japanese TV news deceived people, saying that the rising radiation levels were because of atmospheric pollution coming from China."[240] Japanese political leaders did not seriously undertake programs "to help people understand the situation well enough to make their own behavioral judgments," failing to explain "the risks of radiation exposure to different segments of the population, such as infants and youths, expecting mothers, or people particularly susceptible to the effects of radiation."[241] Taking into account both internal and external communication, "the lack of public disclosure falls as much on the government as it does on TEPCO": "The government's response to the nuclear disaster has also not been fully transparent"; "most communications have been vague, confusing, not particularly timely, and have tended to present a rosier picture than the actual situation"; and "it took a full five hours after the first reactor explosion for the government to report that no radioactive material had been leaked."[242]

In particular, the Japanese government grossly mishandled the evacuation of people from the disaster-stricken area. Although over four hundred thousand people were eventually evacuated, many of those displaced—more than one hundred thousand people—remained homeless even through 2016 because of the continuing government ban on former residents in affected areas returning to their homes.[243] In this mass evacuation, the Japanese government did not even "tell their citizens which direction the massive radioactive cloud was travelling . . . so Fukushima evacuees fled to the north west of Japan—right into the path of the radioactive plume."[244] The government evacuation problems "resulted in thousands of evacuees given an instruction to the direction of [a] more dangerous highly contaminated area, and then again moved to another area, in a few days without realizing what risks they had."[245] The Fukushima Nuclear Accident Independent Investigation Commission found that "the residents' confusion over the evacuation stemmed from the regulators' negligence and failure over the years to implement adequate measures against a nuclear disaster, as well as a lack of action by previous governments and regulators focused on crisis management": only 20 percent of the residents of the town hosting the plant knew about the accident when evacuation was ordered; "many residents had to flee with only the barest necessities and were forced to move multiple times"; "there was great confusion over the evacuation"; "some residents were evacuated to high dosage areas because radiation monitoring information was not provided"; and some of these people were "then neglected, receiving no further evacuation orders until April."[246]

The government-supervised TEPCO had a long history of using deception. In earlier decades, TEPCO had doctored reactor safety records; in 2002, it admitted to falsifying the results of safety tests on reactor 1's containment vessel; in 2003, TEPCO prohibited inspectors from visiting reactors, later acknowledging that it systematically covered up inspection data that showed cracks in reactors; and in 2007, following a fire at another nuclear plant, TEPCO masked the fact that hundreds of gallons of radioactive water leaked out.[247] In the Fukushima crisis, TEPCO was guilty of many deceptive practices—"it consistently failed to inform government officials, especially the prime minister and cabinet, of what was happening at the plant, downplayed dangers, did not adequately inform or misinformed the government and public about developments, and then tried to shift all the blame onto the prime minister rather than admit its epic mismanagement of the crisis."[248] Entrenched corruption facilitated this deceptive pattern, as for example TEPCO "built the reactors on a known fault line and then colluded with government regulators to avoid preparing for the inevitable."[249] TEPCO's ineptitude reportedly "infuriated" Japan's prime minister, who learned of the first plant explosion only from watching television.[250]

Role of Information Overload in Triggering or Changing Ambiguity, Deception, and/or Surprise   When lots of fragmentary pieces of rapidly changing information of uncertain credibility began flowing into the Japanese prime minister's office about the Fukushima disaster as it unfolded, two challenges immediately emerged—accurately integrating and interpreting the security implications of incoming data, and quickly developing policies to minimize the negative repercussions. However, in response to these challenges, "experts seemed incapable of providing any useful guidance":

> With each new report from the site, the prime minister would ask [the experts] what he should do . . . and each time they averted their eyes. . . . Once the prime minister asked one of the scientific experts a question directly, but the expert was at a complete loss for words.[251]

Moreover, when facing the overwhelming flood of technical information, the prime minister "contributed to the confusion" by immersing himself "in the minutiae of the nuclear plant" instead of looking at the big picture and by refusing to delegate authority to make decisions.[252] Compounding these information interpretation barriers under crisis time pressure was the context of "the ingrained conventions of Japanese culture"—reflexive obedience, reluctance to question authority, devotion to "sticking with the program," groupism, and insularity.[253]

Because of the global threat posed by the Fukushima disaster, media coverage of it was extensive, and as a result the mass public was exposed to a barrage of potentially conflicting information from a wide variety of sources. For example, on December 28, 2011, the Turner Radio Network issued an overblown warning that a radiation cloud emanating from Fukushima's reactor 3 posed an imminent three-to-five-day threat to hit the West Coast of the United States, and several other news sources picked up and repeated this ominous claim.[254] The mass public was in no position to judge the veracity of outlandish assertions compared to more prudent ones.

Sometimes access to too much quickly changing information can make matters worse. Rising public concerns about the scope of evacuation areas exemplified this pattern:

> At each stage of the crisis, experts offered [Japanese prime minister Naoto] Kan varying estimates of the area that should be evacuated, and each time, Kan took the safest option. But the estimates kept changing, so Kan kept changing what he told the public. Paradoxically, this emphasis on safety and truthfulness led to growing fear and mistrust. . . . People accused the government of underestimating and playing down the gravity of the situation.[255]

The Japanese government was "repeatedly forced to raise its recommended safety limits for radiation exposure to citizens and workers—otherwise it would have been legally required to evacuate the site immediately"—and as a result "some Japanese people believe that the government is corrupt" and "others think it is incompetent."[256] The pattern that multiple postdisaster radiation readings collected by different agencies "became confusing and lacked credibility" further demonstrated that "when too much information on a disaster is in circulation, it can also have negative effects on public perception of the government."[257]

Rationale and Purpose of Ambiguity, Deception, and/or Surprise   The motive for the Japanese government's unpreparedness and deception surrounding the Fukushima nuclear disaster revolved around wishful thinking, an unwillingness beforehand to consider open-mindedly the dire potential for impending disaster:

Several reasons have been offered to explain why so many in leadership positions ignored the warnings of history. . . . After World War II and the destruction of Nagasaki and Hiroshima by nuclear weapons, the Japanese population vehemently opposed all use of nuclear power in their country. So the government undertook a campaign to persuade people of "the absolute safeness" of nuclear power. . . . Absolute safeness meaning that there was no risk that something could go wrong, no risk that a meltdown could happen. Well, that myth of absolute safeness developed over the years into a culture where it almost became a taboo to even talk about this. . . . Discussing a worst-case scenario was feared because it might bring panic to the citizens. And therefore it was omitted from the regulatory discussions.[258]

In particular, the Japanese government feared "the social chaos caused by such upheaval,"[259] with Japan's nuclear power governance "found overwhelmingly to have long discounted known safety problems of nuclear power."[260] However, Japanese government hesitation went well beyond fear of mass public panic: "Talking about worst-case scenarios was avoided not simply because it would scare people, but because such fear would mean that local communities would oppose the building of reactors, and without local support the reactors would not be built."[261] Although increasing public support was a motive for Japanese state action, in their responses upset Japanese citizens' wanted to show the government that its behavior in the crisis was absolutely unacceptable.

Policy Effectiveness of Ambiguity, Deception, and/or Surprise   The Japanese government's use of deception in the Fukushima disaster toward the Japanese people and the international community was unsuccessful in both its short-run and its long-run strategic goals. Because of the truly devastating impact that ensued, fear, anger, and resentment were rampant among those most affected, radiation dangers continued, and stable security was not quickly restored.

In the wake of the disaster, the Japanese government's attempt to clamp down on undesired information leaking out appeared to worsen both its domestic and its international image. Immediately following the meltdown, a plant employee in Fukushima and his family were "ordered by a local official to stop discussing what they had seen."[262] Furthermore, the Japanese government was "not only under-measuring and ignoring varieties of radioactive threat, but even withholding the iodine pills in 2011 that might have mitigated the growing epidemic of thyroid issues."[263] Later, the Japanese government "passed a harsh state secrets law that threatens to reduce or eliminate reliable information about Fukushima."[264]

Perceived Legitimacy of Ambiguity, Deception, and/or Surprise   The reaction of the international community was shock, yet this was not generally accompanied by appropriate remedial action, perhaps in part because "no risk information or meltdown information were provided [by the Japanese government] during [the] initial weeks"[265] after the disaster. The global credibility and status of the Japanese government took a serious blow, especially given the substantial human life and property damage. Japanese political leaders' tarnished image has been amplified by its shady moves taken long after the end of the immediate crisis. One example of such moves is the "Japanese practice of recruiting homeless people to work at Fukushima in high level radiation areas where someone with something to lose might not be willing to go for minimum wage."[266] Moreover, despite Japanese media proclivities during the crisis to play "the willing government handmaiden in reassuring the public with falsehoods,"[267] another example of questionable moves is the abrupt termination of a three-year-long special "watchdog division" investigation by one of Japan's oldest and most liberal/intellectual newspapers—*Asahi Shimbun*—into the aftereffects of the Fukushima nuclear disaster: "the hastiness of the *Asahi's* retreat raised fresh doubts about whether such watchdog journalism—an inherently risky enterprise that seeks to expose and debunk, and challenge the powerful—is even possible in Japan's big national media, which are deeply tied to the nation's political establishment."[268]

Japanese citizens, normally relatively supportive of government action, reacted to the Fukushima disaster with outrage at the state's response:

> Nevertheless, the *ad hoc*, uncoordinated, non-transparent handling of the crisis, the government's inability to control and manage TEPCO, and the failure to communicate timely and accurate information to the public when it needed it have all resulted in the DPJ [Democratic Party of Japan] government becoming the target to much of the public's criticism. Six months after the disasters, polls showed that 67% do not support the government response to the earthquake and that 78% do not support the government response to the nuclear plant incident.[269]

Although some Japanese citizens fatalistically accepted the disruptive consequences of the Fukushima disaster, the most common mass public reaction was skepticism and resentment toward both government and corporate officials, whose relationship—as is traditional in Japan—was decidedly collusive. The dramatic human security costs were "blamed on corporate greed, poor planning, and compromised government oversight in the years leading up to 11 March 2011."[270] Overall, "Japan's political elite has consistently underwhelmed the public throughout this crisis, wasting too much time and energy on petty politics and failing to act effectively or with sufficient urgency, leaving the public disenchanted to Tokyo and forcing the people . . . to rely on their own initiative."[271]

In addition to popular resentment, mass public fear and feelings of helplessness were widespread throughout Japan—not just in areas close to Fukushima—because of the uncertainty about long-term nuclear radiation effects, particularly poignant "for a country that experienced two nuclear bomb attacks in 1945."[272] The perceived long-term dangers were central in the minds of Japanese citizens—"fear and uncertainty among the population is focused not on an event in the horrific recent past (the tsunami), but more on one in the undisclosed future—a time to come when today's children may find themselves victims of disease, social stigma, or both."[273] Because of the Fukushima disaster, "in Japan public anxiety about nuclear power was at an all-time high, informed by distrust and suspicion of government and industry about the exact reasons for Fukushima, apart from the obvious contribution of natural ones."[274] Much of the mass public was aware that "any type of nuclear emergency requires immediate action in an effort to protect citizens"; although government authorities "typically administer high doses of potassium iodide to those affected in an effort to block thyroid absorption of iodine 131," people felt the "need for better, more effective protective options."[275] Thus, for months following the Fukushima disaster, it "seemed like an endless national nightmare."[276]

Because of state policy flip-flopping, the Fukushima disaster ended up seriously undermining Japanese citizens' faith in the Japanese government's ability to provide them with their personal security needs:

> Once the first week had passed, the government created the impression that things were calming down in the disaster zone, and people started to return home. . . . Then the government started to lose confidence in its risk-analysis, and ordered fresh evacuations. . . . Meanwhile many people lost faith in the government and made their own evacuation arrangements. Most of those who left have still not come back. The population of Fukushima prefecture fell by about 75,000, from 2.024 million to 1.950 million, over the two years following the disaster.[277]

Even considering the later follow-up contrition and admissions of culpability by government and corporate officials, popular trust in the ruling regime was not restored, and more citizens felt on their own for disaster protection. More broadly, Japanese government security officials may find themselves now trapped in what some call "the paradox of safety"—they feel "compelled to justify an inherently dangerous technology, they promise absolute safety, then when any failure occurs, they find that the public, sensitized as much as reassured by their past explanations, loses all trust."[278]

Unintended Consequences of Ambiguity, Deception, and/or Surprise   The Fukushima disaster had major long-term negative effects on human health. Residents in the affected area have struggled with its aftereffects—"they continue to face grave concerns, including the health effects of radiation exposure, displacement, the dissolution of families, disruption of their lives and lifestyles and the contamination of vast areas of the environment," with "no foreseeable end to the decontamination and restoration activities that are essential for rebuilding communities."[279] Indeed, "the Japanese were kept in the dark from the start of the Fukushima disaster about high radiation levels and their dangers to health," and "in order to proclaim the Fukushima area 'safe,' the government increased exposure limits to twenty times the international norm."[280] In June 2011, Greenpeace International executive director Dr. Kumi Naidoo unveiled a sample of contaminated soil from a Fukushima playground at a Tokyo press conference while criticizing the Japanese government's response to the Fukushima nuclear crisis and its ongoing failure to prioritize protection of the health and welfare of its people:

> "On Tuesday I had the privilege of meeting with both teachers and school children in Fukushima. Looking at innocent faces of the children, I found it difficult to contemplate the dangerously high levels of radiation they are being exposed to on a daily basis," said Dr. Naidoo. "While the Japanese government has made some efforts at decontamination, it is not nearly enough. These are real people we are dealing with here, not statistics." . . . "The pregnant woman [*sic*] and children of Fukushima are the future of this country. They are innocent victims of Japan's insistence on using nuclear energy and there is a moral imperative for them to be evacuated from high-risk areas until proper decontamination is carried out."[281]

Although "the mainstream media abruptly stopped reporting on it just a few months after it began," as of March 2013—two years after the Fukushima disaster—the situation at the Daiichi nuclear plant remained "critical."[282] In June

2014, the ruins of Fukushima were "still smoldering" and the negative health consequences were "more pronounced than ever," with the continuing health risks including skin contamination, psychological trauma, cancer, thyroid damage and thyroid cancer in children, and pregnancy issues in women.[283] Overall, the "recovery is slow and may take decades,"[284] as removing the melted nuclear fuel from the reactors at the Fukushima Daiichi nuclear power plant has not yet begun, and "decommissioning of the nuclear power plant is said to be a process that will take 30 to 40 years."[285]

The Fukushima disaster also triggered great interest in renewable energy[286] and increased concerns about relying on nuclear power. In the immediate wake of the Fukushima nuclear power plant shutdown, the Japanese government announced several wind and solar projects, and it "has set renewable targets of between 25% and 35% of total power generation by 2030, by which time some $700 billion would be invested in new, renewable energy."[287] The widespread local fears of nuclear contamination engendered "not only a breakdown of trust in the government but also countless more complex divisions among citizens"—now Fukushima-area farmers are unable to market their crops to fellow citizens because of the perceived contamination risks.[288] Globally, although the United States did little to reverse its nuclear power commitment despite calling for greater safety precautions, Germany made a dramatic about-face: shortly after German chancellor Angela Merkel had hatched a plan to keep the country's seventeen plants open for an additional twelve years, post-Fukushima German citizen antinuclear protests caused her to reverse course, and now all of Germany's nuclear plants will be shut down after 2022.[289]

A final unintended consequence of the Fukushima disaster was increased distrust of Japan by its allies and the international community, who had also been deceived about the crisis. Indeed, "it was literally impossible for the world community to get a clear understanding of, and the truth about, the Fukushima nuclear disaster," for it was clear that there was "no trustworthy source of information."[290] Even between Japan and the United States "there were some mutual misunderstandings seen in the initial management of information"[291] during the Fukushima crisis.

Lessons for Future Management of Ambiguity, Deception, and/or Surprise   One Fukushima lesson is that concerns about preventing mass public panic can go too far if a state deceptively withholds vital disaster information or declares a crisis to be over when human security is still in jeopardy. Just as the American government has been reluctant to inform its citizens of how devastating a nuclear bomb attack would actually be, so the Japanese government feared telling its citizens about the true extent of the dangers emanating from the Fukushima radioactive fallout. With the stakes as high as they were in the disaster's aftermath, the affected public had a right to know at least basic information about what the dangers were and how to protect themselves. If a national government does not provide such information and massive death and property damage ensues, then citizen trust in the state declines precipitously.

A second lesson is that when public and private deception is mixed with public and private corruption, the result can be toxic in terms of public trust and global credibility. The Japanese government's collusive relationship with the corrupt and deceptive TEPCO played a major role in causing the Japanese people to lose faith in the state's potential to equitably and effectively manage the aftermath of the crisis. Such faith is extremely difficult to restore.

A third lesson is that internal and external deception seems most dangerous when people's lives and property are at stake. When onlookers suspect deception initiators of intentionally withholding or distorting information that could save their lives, the subsequent resentment and anger seem likely to be particularly intense. This risk seems especially high if some of those affected believe that initiators are themselves responsible for mismanagement in a way that led to the enlargement of the risk, as is the case with the Fukushima disaster.

A fourth lesson is that a boatload of incoming information is hard to assimilate when it involves changing estimates and warnings over time. With Fukushima, such adjustments were viewed as a sign of state incompetence rather than state attentiveness and adaptability. The most likely security outcome of this kind of fluctuation seems to be mass confusion.

2008 Russian Invasion of Georgia

Long-standing tensions within Georgia have existed since the early 1990s, with two areas—Abkhazia and South Ossetia—highly partial to Russian rule, and with most of their residents granted Russian citizenship and passports.[292] When Mikheil Saakashvili was elected Georgian president in 2004, he pledged to institute democratic reforms and to regain control over the separatist regions. In September and October 2006, Georgia expelled six Russian intelligence agents accused of espionage, and Russia responded with a full economic embargo of Georgia, severing all transportation and communication links. In July 2008, the Georgians and the South Ossetians exchanged artillery fire, and Russia initiated a large-scale military exercise nearby.[293]

In early August 2008, these tensions escalated between Russia-leaning South Ossetia and more democratically oriented Georgia, with each accusing the other of launching hostile artillery barrages, and on August 8, "as world leaders gathered in Beijing to watch the opening ceremony of the Olympic Games, Russian tanks rolled across the border into Georgia."[294] Russia also launched massive air attacks throughout Georgia, and Russian troops engaged Georgian forces in South Ossetia. Then Russian warships landed troops in Georgia's Abkhazia region and took up positions off Georgia's Black Sea coast. Georgia claimed that South Ossetian forces did not respond to a cease-fire appeal but intensified their shelling, "forcing" Georgia to send in troops, while Russia claimed that its forces entered the area simply as a "peacekeeping" operation. On August 26, Russia officially recognized the independence of the two breakaway regions of Abkhazia and South Ossetia. As a result of the Russia-Georgia conflict, nearly 1,000 people died,[295] and by October 2008 the World Bank reported that about 127,000 people were displaced by the fighting in Georgia, South Ossetia, and Abkhazia.[296]

Description of Ambiguity, Deception, and/or Surprise   Russia used strategic deception to conceal the true levels of planning and offensive intention surrounding its military action in Georgia:

> The deception model might be described as follows. Russian leadership . . . made calculated decisions in their response to the Georgian attack to force a specific Western perception that they were only responding to aggression. . . . Thus, leaders wished to convey the notion that they were forced to respond to actions carried out by Georgia, and that little forethought went into planning a response; this explanation places emphasis on Russia as the defenders of South Ossetia and Abkhazia. . . . Ideally, this would then lead to a misperception and false assessment of Moscow's intentions.[297]

The desired payoff from such deception was enhancing the perceived legitimacy of Russian actions—creating "the illusion that Moscow was not prepared for war and was taken by surprise" could maximize the chances of "masking the notion that Russia had been prepared to respond, based on months of incursions from both sides prior to the war" and minimize the chances of inducing "a wider response, possibly with NATO involvement."[298]

Russia also intentionally engendered strategic surprise in its invasion of Georgia. The result was "the world's stunned astonishment over a broadly predictable, thoroughly old-fashioned reaffirmation of big stick intervention, meant to demonstrate the renewed vigor of a long-humiliated, now-prospering great power dominated by an ice-blooded, fiercely ambitious leader, i.e., someone very like Vladimir Putin."[299] In particular, Western states were "aghast at the expansion of Russia's initial supposedly 'defensive' military foray in South Ossetia into a wider occupation of numerous regions of the Georgian state and at the Kremlin's open repudiation of Georgia's territorial integrity."[300] Although "the world's political elite" were "caught off guard" by this development, some outside analysts felt this Russian response to instability within its sphere of influence should have been highly predictable: "from ancient times to the present, great powers have jealously sought to enforce their regional supremacy and expel rivals from their neighborhood—America's Monroe Doctrine, proclaimed in 1823, conformed to this practice by unilaterally decreeing that the Western Hemisphere was no longer subject to further European colonization or interference."[301] From this cynical perspective, "it was but a matter of time before traditional Russian concerns and strategies resurfaced after the Soviet Union dissolved," and "given Washington's embrace of Saakashvili, at some point there was bound to be a robust Russian counterstroke," at least partially as a warning to other former Soviet republics.[302]

Role of Information Overload in Triggering or Changing Ambiguity, Deception, and/or Surprise   Information overload was tangibly present regarding the Russian invasion of Georgia. Almost as soon as the Russian intervention began, "the Georgian government began sending hourly e-mail updates to foreign journalists."[303] In response the Western media appeared to have an insatiable appetite for coverage of the war:

> From the start of the conflict between Georgia and Russia, the leading Western news agencies dedicated several reports daily to this conflict. The desire to understand the events was overwhelming. Experts on Russia and the Caucasus were in high demand by local and national news agencies. Initial assessments rapidly found their way into major American and European news channels and newspapers.[304]

Within the flood of information, two "contrary narratives" have stood out: the Russians claimed that their actions in Georgia have been "essentially retaliatory—an ad hoc, exceptional, though large-scale, response to a Georgian attack in South Ossetia"—while Russian critics have argued that "Russia's ostensible commitment to protect 'Russian citizens,' a core justification of the intervention in Georgia, has principally served as a means of coercion and a device to expedite military intervention in that country for other strategic purposes."[305] These diametrically opposed Western and Russian interpretations of the crisis have provided the basis for global consternation about what was really transpiring.

Partly because of this noisy information environment, misperception surrounded the conflict between Russia and Georgia. In terms of Georgian misperceptions, "Georgia acted irresponsibly in attempting to reconquer the secessionist region of South Ossetia by force."[306] Moreover, Georgia's expectation of tangible Western support in its conflict with Russia was decidedly misplaced: "Georgia's macho president mistook expressions of American solidarity for a serious commitment to intervene, and leaped at the bait by overreacting—just as the Kremlin hoped he would, when Russia's so-called peacekeepers stood by as loutish local militias misbehaved within and beyond the borders of South Ossetia."[307] In terms of Russian misperceptions, "Russia acted irresponsibly in militarily intervening to prevent Georgia from overrunning South Ossetian militias and Russian peacekeepers."[308] Notably, "all sides justified their actions in terms of international law, with Georgia arguing that it acted to prevent violent secession, and Russia arguing that it was protecting ethnic Ossetians and Russian peacekeepers from indiscriminate, even genocidal, attacks."[309]

Within this confusing information setting, in which every side is able to paint its actions as justified, manipulation possibilities abounded, and Russia especially took advantage of them. One example of such manipulation, reflecting Orwellian "doublespeak," was that Russia was able successfully to mask Soviet-style strategy and tactics by relabeling "aggression" as "peacekeeping."[310] A second manipulation example was that Russian propaganda was able to offer up "absurd" explanations of the true reasons for American involvement in the conflict: "*Vesti FM*, an outlet of Russian state radio, has been broadcasting that Vice President Cheney set these events in motion in order to prevent the election of Senator Obama."[311] A third example was a sophisticated disinformation campaign at work—Russia accused Georgian forces of ethnic cleaning and "widespread human rights violation in South Ossetia," a contention refuted by independent human rights organizations and in any case outweighed by Russian endorsement of systematic ethnic cleansing of ethnic Georgians from South Ossetia.[312]

Nonetheless, in attempting to manipulate information externally to promote positive global reaction, Russia lost the propaganda battle. Georgia launched its own skilled propaganda campaign, attempting to curry Western support and dominating international headlines:

The English-speaking, Columbia University–educated Georgian president, Mikheil Saakashvili, appeared live on CNN. In subsequent interviews and speeches, he hit every major talking point meaningful to Western audiences, including claims of ethnic cleansing and genocide—and the bizarre allegation that Russia was plotting to start forest fires. Meanwhile, his government stage-managed rallies featuring EU flags and called for Europe to rescue the embattled democracy.[313]

Meanwhile, Russia seemed to lack an effective international public relations strategy:

> By contrast, Russia's public relations effort was feeble. Images of hapless Ossetian refugees clogged Russian television screens, but Moscow made few attempts (beyond awkward press briefings by a uniformed general and a benefit concert in the bombed-out ruins of Tskhinvali) to impress its version of events on the international media.[314]

Perhaps the Russian premise behind this weak external information manipulation effort was that with superior military force one does not need to care about international image.

Rationale and Purpose of Ambiguity, Deception, and/or Surprise   Russia had multiple aims in using strategic surprise and deception in the Georgia invasion. Russia's long-term strategic goals included "increasing its control of the Caucasus, especially over strategic energy pipelines" and "recreating a sphere of influence (a 'sphere of privileged interests' in official Russian parlance) in the former Soviet Union and beyond."[315] Russia saw the United States as the key instigator of the crisis: even though the Bush administration had previously "cautioned Georgia against actions that might result in a Russian military response,"[316] "to Prime Minister Vladimir Putin, the true culprits were most probably the Americans who encouraged Georgia's unprovoked attack on breakaway South Ossetia."[317] Thus Moscow "viewed as affronts" American president George W. Bush's emboldening of "Georgia's defiance of Moscow by training its armed forces and favoring its fast-track accession to NATO" and visiting Tbilisi to partake in "a rhapsodic rally, thereby demonstrating his personal support for a pro-Western, American-schooled Georgian president."[318]

Georgia's primary goals in the conflict were to reinforce its sovereignty, territorial integrity, and independence from Russia and to move closer to membership in NATO and to institution of democratic practices.[319] A January 2008 national poll revealed that most Georgian citizens favored NATO member-ship.[320] From Georgia's perspective, the two culprits in the conflict were "thug-gish South Ossetian militias, acting in collusion with Russia," which "precipi-tated the conflict by firing on Georgian villagers and peacekeepers on August 6 in violation of a 1994 armistice agreement."[321]

Policy Effectiveness of Ambiguity, Deception, and/or Surprise   From a strictly military perspective, the Russian use of force in Georgia was successful in achieving its short-run goals. Russian military action in Georgia consistently "was well-thought-out and properly resourced, giving Russia significant advan-tages at the operational level of war," and "at the strategic level, Russia was able to execute a combined political-military strategy that isolated Georgia from its western partners while setting the conditions for military success."[322] Indeed, "the swiftness with which large Russian contingents were deployed after 8 Au-gust into South Ossetia and beyond was remarkable."[323] To minimize chances of future Georgian belligerence, Russia "did its best" to achieve "the destruc-tion or seizure of Georgian army, air force and naval military equipment and infrastructure, and the targeting of all Georgian military facilities and bases."[324] By the end of the five-day war, "after the guns of August 2008 fell silent, it was clear that Abkhazia and South Ossetia would never return to full Georgian control."[325] However, the long-run broader political consequences of Russia's invasion of Georgia seem much more uncertain.

Perceived Legitimacy of Ambiguity, Deception, and/or Surprise   In terms of legitimacy, the West totally sided with Georgia, strongly criticizing Russia's "display of brutality."[326] Western legal sources claimed that the Russian invasion of Georgia violated both the UN Charter and customary international law:[327]

> U.S. and European leaders immediately condemned Moscow for flouting estab-lished borders. Tired Cold War metaphors—of containing the bear before the next domino fell—reappeared with startling rapidity. The Western press painted President Dmitry Medvedev and Prime Minister Vladimir Putin as leaders of a rogue government scheming to roll back democracy and monopolize oil and gas networks across Eurasia.[328]

Western voices characterized Vladimir Putin, who was still considered to be in charge even when Medvedev was president, as having "disregarded the rule of law, rigging the political process in favor of his personal rule, virtu-ally extinguishing freedom of speech, and bringing the energy industry under state control" through "reliance upon force," which "unleashes corruption and crime."[329] The American government specifically condemned Moscow's unilateral recognition of breakaway provinces South Ossetia and Abkhazia as sovereign entities.[330]

However, despite Western verbal disapproval of the Russian invasion, few serious action commitments followed, disappointing Georgia. Instead, "the Western response has been to mollify and placate," with the Europeans being "outright appeasers":[331] "There is little stomach in Brussels or Washington for trying to roll back the independence of Abkhazia and South Ossetia, or for making Russia's troop presence there a central issue in relations with Moscow."[332]

Meanwhile, Russia enjoyed overwhelming domestic approval for the Georgian invasion. A public opinion poll by the respected Moscow-based Levada Center reported that "almost 80 percent of the Russian respondents approved of it," and "over half blamed Georgia for initiating the conflict and identified the United States' desire for influence in the Caucasus and the greater Black Sea region as the root cause."[333] As to legitimacy, "a military operation that the West denounced as an act of aggression was seen in Russia and beyond as laudable, proportionate, and humanitarian."[334]

Unintended Consequences of Ambiguity, Deception, and/or Surprise   A key unintended consequence of Russia's invasion of Georgia was the intensification of Russia's political isolation. Despite extensive Russian lobbying, "as of October 2008 no CIS [Commonwealth of Independent States] state had been persuaded to give formal recognition to South Ossetia or Abkhazia."[335] This widespread reluctance "provides graphic evidence of Russia's international isolation on the issue and the risks that former Soviet republics associate with redrawing post-Soviet international borders, especially if the dominant regional power is driving the process and perhaps even expecting to acquire new territory as a result."[336]

Another important unintended consequence of the Russia-Georgia conflict is that the former Soviet satellite states—governments in the Baltic States, Azerbaijan, Ukraine, Moldova, and Central Asia—realized that "Russia has effectively nullified its credibility to serve as an honest broker in conflicts on or near their territory" and that "the West is unlikely to step forward to defend its own and their interests in the case of military aggression."[337] In other words, both the United States and Russia are now hamstrung in promoting peaceful transformation in the region.

Lessons for Future Management of Ambiguity, Deception, and/or Surprise   One lesson from the Russian invasion of Georgia is that, despite claims of the spread of enlightened liberal global norms, "might makes right" still is alive and well internationally:

> In the manner of the Soviet Union, Russia is taking it upon itself unilaterally to decide what the boundaries of other countries are to be and who is to govern them. Consternation grips neighbors who have been through this life-and-death experience before, within living memory. The Baltic republics, the Asian Muslim republics, and Moldova all contain ethnic-Russian inhabitants, on whose behalf some grievance is easy to concoct, allowing the Kremlin to treat these neighbors in the same way it has treated Georgia.[338]

History has taught Russia to operate on the basis of narrow power-maximizing self-interest:

> Russia eventually traded its partnership with Europe for a wary cynicism, an introverted nationalism, and a belief in raw power as the hallmark of international politics. . . . The difference today is that there are plenty of other countries, from China and Venezuela to Iran and Syria, that share Russia's view of the global order. And there are others, such as India and Turkey, that at least understand it. Russia is not alone in questioning the consistency of the United States' responses to territorial conflicts around the world or the evenhandedness with which the West doles out labels such as "democratic," "terrorist," or "rogue state."[339]

Russia consequently has "embarked on a new era of muscular intervention," one in which it tends "to see hard power as the true currency of international relations,"[340] providing "proof that the age of conventional warfare is far from over."[341] Within a world where information manipulation is so easy that almost any concocted excuse can be credibly projected as an excuse for aggression, such anarchic behavior is decidedly not out of place. Specifically, "Russia's insistence on its right to defend by force its citizens outside its borders is open to manipulation," especially "when a country first confers its citizenship on a large number of people outside its borders and then claims it is entitled to intervene coercively to protect them."[342] Thus many analysts now believe that in the future "the West now can expect that a nationalist Russia will mobilize its resources to probe every weak point around its borders."[343]

A second closely related lesson is that intergovernmental organizations' power in today's world may be dramatically overstated. Through invading

Georgia, Russia showed "little faith in multilateral institutions, such as the UN Security Council or the Organization for Security and Cooperation in Europe, in which it exerts considerable influence," at least in part because "Russian leaders believe that the existing multilateral institutions are unsubtle fronts for promoting the naked interests of the United States and its major European allies."[344] The Russian invasion specifically "eroded the effectiveness of the NATO umbrella in Eastern Europe, even though Georgia is not yet formally a member, since it became apparent that Moscow can use force against its neighbors with relative impunity."[345] More broadly, the war "demonstrated the weaknesses of NATO and the EU security system, because they provided no efficient response to Russia's forced changing of the borders and occupation of an Organization for Security and Cooperation in Europe (OSCE) member state."[346] The regional security consequences of "an emboldened and mistrustful Russia" has made "the future of NATO uncertain," "left the United States and its allies divided over Moscow's role in the world," and "laid bare the United States' inability to deter friends from behaving like fools."[347]

A third lesson underscores the realization that winning the battle to manipulate external information in such a way as to garner global support and sympathy does not necessarily translate into foreign security policy success. Georgia won the external propaganda battle, but it lost the war and did not receive the tangible support from the West that it wanted and needed. Nonetheless, "in the future, the real contest will be over which powers are best able to spin their flaws and speak convincingly to an increasingly savvy world citizenry that is as skeptical about the United States' messianic democratizing as it is about Russia's nationalist posturing."[348]

Strategic Surprise Cases

2007 Israeli Attack on the Syrian Al-Kibar Nuclear Plant

On September 6, 2007, in an attack shortly after midnight, seven Israeli Air Force fighter jets executed Operation Orchard and—having electronically blinded Syria's air defense system—destroyed a covert site of a future nuclear reactor near al-Kibar in northeast Syria. The planning for this violent act had begun months earlier: in mid-May 2007, Prime Minister Ehud Olmert contacted American government officials asking if Mossad chief Meir Dagan could share intelligence that Syria was building a nuclear reactor, with North Korea providing both the design and the technical assistance.[349] After the attack, follow-up investigations revealed that Namchongang Trading, a North Korean firm, had smuggled in key materials for the Syrian reactor from China and possibly Europe.[350] The CIA later revealed that the Syrian plant was indeed a nearly completed gas-cooled, graphite-moderated nuclear reactor secretly under construction since 2001.[351] The unit may have been "part of an Iranian-led multinational nuclear weapons effort, with both Syria and North Korea being collaborators.[352] Given the location selected and the absence of an electrical grid connection, it was clear that this reactor was part of a nuclear weapons program rather than electric power production.[353] Notably, secretly constructing a nuclear reactor would constitute "a violation of Syria's obligations under the Nuclear Non-Proliferation Treaty."[354]

On April 24, 2008, the CIA released a video showing that the Syrian facility had concealed construction of a nuclear reactor similar to North Korea's Yongbyon reactor. In April 2011, the International Atomic Energy Agency (IAEA) confirmed that the Syrian target was indeed a nuclear reactor site.[355] ABC News reported that, prior to the airstrike, Israel had recruited a spy to take ground photographs of the nuclear reactor's construction from inside the complex.[356] After the destruction of the facility, "Syria moved quickly to cover up its covert nuclear activities by demolishing and burying the reactor building and by removing incriminating equipment";[357] soon after the bombing, the Syrian government bulldozed the reactor site, and when a 2008 IAEA site visit found uranium traces, Syria never permitted a return visit.[358] Despite frequent opportunities to do so, the Syrian government was never willing at any point later to admit details about the true nature of this facility.

Description of Ambiguity, Deception, and/or Surprise   The Israeli attack on the Syrian nuclear plant was an unambiguous application of strategic surprise. This surprise was designed to outflank Syrian strategic ambiguity and surprise in building the secret nuclear plant. An American deputy national security advisor reported that Syria building a nuclear reactor was unexpected to both Israel and the United States:

> Initially, there were doubts that Bashar al-Assad could be so stupid as to try this stunt of building a nuclear reactor with North Korean help. Did he really think he would get away with it—that Israel would permit it? But he nearly did; had the reactor been activated, striking it militarily could have strewn radioactive material into the wind and into the nearby Euphrates River, which was the reactor's source of water needed for cooling. When we found out about the reactor, it was at an advanced construction stage, just a few months from being "hot."[359]

To maintain the advantage of an unexpected airstrike, the Israeli operation had to keep everything close to the vest from the moment joint American-Israeli planning began in May 2007:

> This process was run entirely out of the White House, with extremely limited participation to maintain secrecy. The effort at secrecy succeeded and there were no leaks—an amazing feat in Washington, especially when the information being held so tightly was as startling and sexy as this.[360]

This is certainly one of those foreign security policy cases in which, if even the slightest word of the impeding military attack had been disclosed in advance, the entire operation would have had to be scuttled.

Role of Information Overload in Triggering or Changing Ambiguity, Deception, and/or Surprise   Following the attack, a glut of information emerged through international media about the Israeli airstrike on the Syrian al-Kibar nuclear plant. Ironically, "despite official silence in Tel Aviv (and in Washington), in the days after the bombing the American and European media were flooded with reports, primarily based on information from anonymous government sources, claiming that Israel had destroyed a nascent nuclear reactor that was secretly being assembled in Syria, with the help of North Korea."[361] So a seemingly contradictory but ultimately effective information dissemination strategy was evident here from the strategic surprise initiators—"it was evident that officials in Israel and the United States, although unwilling to be quoted, were eager for the news media to write about the bombing."[362]

However, despite this plethora of news coverage, interpretive agreement about what actually transpired and understanding of the events surrounding the attack remained low:

> Much remains unknown about the September 6 incident in which Israeli warplanes entered Syrian air space. Not only do commentators disagree about the various motives of the diverse participants involved, but even the basic facts remain in dispute. Neither the Israeli, Syrian, nor U.S. government has offered a detailed description of what occurred. Outside experts and media commentators have filled the data vacuum by offering their own diverse interpretations about what precisely happened that night.[363]

As is typical of information overload's impact when it incorporates multiple interpretations of a significant foreign security action, "the mystery surrounding Israel's apparent air strike against Syria on September 6 gave observers consid-

erable leeway to interpret the ambiguous event—Syrian leaders describe the affair as an Israeli stratagem designed to bolster the credibility of Israel's discredited military deterrent or disrupt unwelcome peace initiatives in the Middle
East," while "in contrast most Western media coverage implies that the target
either involved a shipment of nuclear technology from North Korea or some
other object of proliferation concern that alarmed Israeli officials sufficiently
that they felt compelled to act to counter a genuine threat."[364] Because of the
extreme secrecy shrouding the Israeli airstrike, confusion persisted about what
really transpired in such a way that allowed considerable spin based on preconceived bias to play a major role in interpreting the event. Moreover, given the
history of unending intense hostility between Israel and Syria, it appears fair to
assume that Syria in the past has regularly perceived numerous signals of impeding aggressive Israeli military action toward it, and that at least occasionally
Syria has had difficulties separating meaningful signals from the noise, sorting
the serious threats from the idle chatter.

Rationale and Purpose of Ambiguity, Deception, and/or Surprise   The motivation for the Israeli use of strategic military surprise was deterrence—a
senior Israeli official stated that it was designed to "re-establish the credibility of our deterrent power," signaling that Israel had zero tolerance for even
a potential Syrian nuclear weapons program.[365] Beyond preventing the Syrians from developing a weapon of mass destruction usable against it, Israel
wished to reinforce deterrence broadly in the Arab world: "Israel's failures in
its 2006 war with Hezbollah weakened the perceived deterrent that it held over
its neighbors; the al-Kibar strike may have been an attempt to reestablish the
supremacy of Israel's military apparatus in its enemies' eyes."[366] In every respect, this action represented classic preemptive military action to eliminate
a potential major threat. Israel could not afford to wait to act until the reactor was fully functional because doing so would risk spreading contaminated
radioactive material.[367] Israel's greatest fear has been getting a hostile "nuclear
neighbor," "particularly one with which Israel has been in a constant state of
war since the Jewish state's independence in 1948"; it is even possible that the
attack on al-Kibar may have served as a warning to Iran or as a practice run for
a future raid on an Iranian nuclear site.[368] Ultimately, "Syria had crossed what
the Israelis regarded as the 'red line' on the path to building a bomb, and had to
be stopped."[369] Because Syria never acknowledged its nuclear plant construction, it is impossible to definitively pinpoint its rationale for doing so at this
particular time.

Israel had considered diplomatic options prior to the airstrike, with the idea that if they failed then the military option could be executed: these included informing the IAEA of the situation, demanding immediate inspections and cessation of work on the reactor, and then—if Syria refused—going to the UN Security Council and demanding action.[370] However, diplomacy was quickly rejected, as Israel distrusted the United Nations, believed that Syria's allies in the United Nations—especially Russia—would protect it, and knew that the IAEA had a leaning against meaningful sanctions in such cases.[371] Moreover, use of force as a last resort after the failure of diplomacy made little sense in this case:

> The argument that there would always remain a military option as a last resort was misleading at best. Once we made public our knowledge of the site, Syria could put a kindergarten right next to it or take some similar move using human shields. Military action required secrecy, and once we made any kind of public statement about al-Kibar, that option would be gone.[372]

Thus a military strike destroying the nuclear site seemed to Israel to be the only viable option.

Policy Effectiveness of Ambiguity, Deception, and/or Surprise   The immediate outcome of this airstrike was military success—Syria could not develop nuclear weapons, and a key threat potential was eliminated.[373] The London *Spectator* announced that the Israeli attack "may have saved the world from a devastating threat."[374] Moreover, Syria has made no clear nuclear move since that time—"there is nothing to suggest that Damascus will or is even able to play with fire once again."[375] Diplomatically, it seems that the Israeli attack "made the Syrians more, not less, desirous of talking to the Israelis because it made them afraid of Israeli power."[376] Furthermore, the attack thwarted any plans Iran may have had to build a backup plutonium facility in Syria,[377] and the airstrike sent a clear message to Iran that "America and Israel can identify nuclear targets and penetrate air defenses to destroy them."[378] Notably, leading up to the Israeli airstrike, the "intelligence collection, and the detection of the Syrian nuclear reactor, represented an unqualified intelligence success."[379] So Israel accomplished its deterrence purpose. In the long run, as the years since the attack have passed, Israeli-Syrian hostile relations have shown no sign of thawing, but so far Syria has initiated no new plan to use force against Israel.

Perceived Legitimacy of Ambiguity, Deception, and/or Surprise   The global verbal reactions to the Israeli airstrike were decidedly underwhelming:

What was particularly notable about this attack was what occurred afterward: the near total lack of international comment or criticism of Israel's action. The lack of reaction contrasted starkly to the international outcry that followed Israel's preventive strike in 1981 that destroyed Iraq's Osiraq reactor. To be sure, foreign governments may have reserved comment because of the lack of information after the attack. The Israeli and U.S. governments imposed virtually total news blackouts immediately after the raid that held for seven months, and Syria was initially silent on the matter and then subsequently denied that the bombed target was a nuclear facility. Yet, the international silence continued even after the CIA on April 24, 2008, provided a 12-minute video and an extensive briefing that made a strong case that the target was a North Korean–built reactor designed for producing weapons-usable plutonium.[380]

A clear act of war and violation of territorial sovereignty—which because of the "absence of a threat of imminent attack from Syria . . . was not a lawful exercise of anticipatory self-defense"[381]—failed to elicit even a murmur of international organization disapproval, and the matter was never raised in the UN Security Council or the General Assembly.[382] In the end, "this lack of international criticism was remarkable, particularly given that past military action by Israel has invariably been the subject of rigorous scrutiny by the international community."[383]

Stunningly, after the strike, "no Arab government commented on the Israeli raid, much less pressed for retaliation against Israel, diplomatic or otherwise."[384] Right afterward, "Syria denounced Israel for invading its airspace, but its public statements were incomplete and contradictory—thus adding to the mystery."[385] Through today Syria and Israel have both "largely adhered to a bizarre policy of downplaying what was clearly an act of war."[386] Even well over a year later, Syria remained temperate in its comments on the incident:

"The facility that was bombed was not a nuclear plant, but rather a conventional military installation," Syrian President Bashar Assad insisted . . . at his palace near Damascus in mid-January 2009. "We could have struck back. But should we really allow ourselves to be provoked into a war? Then we would have walked into an Israeli trap."[387]

Because Israel had accurately calculated before the airstrike that "as long as Assad could deny the existence of the reactor, he would not feel pressured to retaliate," Israelis "helped secure that zone of denial" in that they "briefed their regional allies, including Egypt and Jordan, and urged their leaders to refrain from making public statements about the strike."[388]

What little criticism of Israel emerged came mainly from the Egyptian director general of the IAEA, Mohamed ElBaradei, who learned of the Israeli attack only from media reports. He forcefully condemned both the United States and Israel for their "shoot first and ask questions later" strategy,[389] vehemently contended that the Israeli action was "a violation of international law," and asserted that if Israelis and the Americans had intelligence about an illegal nuclear facility, the IAEA should have been immediately notified.[390] He even went on to claim that IAEA experts "who have carefully analyzed the satellite imagery say it is unlikely that this building was a nuclear facility."[391] However, during the IAEA's own visit in June 2008, "it discovered microscopic uranium particles at the site of the destroyed facility; although the small size of the particles made extensive analysis difficult, the IAEA indicated that the particles consisted of chemically processed uranium, raising concerns that the site had some nuclear purpose."[392] Moreover, in late April 2011, subsequent IAEA director general Yukiya Amano explicitly confirmed that the facility destroyed by Israel was a nuclear reactor under construction.[393]

Considering its complicity in planning the strike, after initial hesitation the United States was supportive of the Israeli force use, believing that the Syrian reactor "was not intended for peaceful activities" and asserting that "the Syrian regime must come clean before the world regarding its illicit nuclear activities."[394] Given that "the mission emerged from more than two decades of comprehensive intelligence collection and analysis by Israeli and American intelligence services targeting Syria's development of weapons of mass destruction,"[395] the Israeli strike was in the end "a model both of U.S.-Israel collaboration and of interagency cooperation without leaks."[396] The United States had itself considered launching an attack on the Syrian nuclear facility, but President George W. Bush had decided that "bombing a sovereign country with no warning or announced justification would create severe blowback," adding that such a covert attack would be too risky.[397]

Even though in the airstrike's "immediate aftermath nothing was heard from the government of Israel,"[398] thanks to awareness via numerous fragmentary media reports Israeli citizens viewed this action very positively. The BBC reported that "the apparent strike on the reactor, deep inside Syria, was seen by many in Israel as a sign of their military prowess."[399] The global image of high-quality Israeli intelligence was decidedly reinforced. Moreover, Prime Minister Olmert's personal domestic popularity received a boost, as an Israeli newspaper poll immediately after the attack showed a 10 percent jump in his public approval rating, with Israelis backing the air raid by nearly an 8-to-1 ratio.[400] However, because of pervasive public fear of foreign attack in Israel from its persistently hostile neighbors, no sense of military complacency emerged afterward.

Unintended Consequences of Ambiguity, Deception, and/or Surprise   Three major negative unintended repercussions remained at the time potential Israeli concerns. First, the Israeli attack on the Syrian nuclear power plant could have inadvertently encouraged illegal nuclear weapons facilities to be built in much harder-to-detect locations, such as the secure underground tunnels where the centrifuges Iran is using for uranium enrichment are now located;[401] this dispersed subterranean positioning of nuclear facilities could make any future foreign demolition operation much more complex, creating much greater Israeli vulnerability.[402] Second, if the Israeli airstrike had hoped to uncover Iran's air defense weaknesses, the strategy may have backfired, because the failure in the Syrian air defense system apparently prompted Iran in December 2007 to purchase the more advanced Russian S-300 air defense system.[403] Third, the possibility of foreign retaliation for the Israeli attack was not completely eliminated: Hezbollah commander Imad Mughniyah, a "notorious terrorist mastermind," reportedly had planned to avenge the Israeli strike on al-Kibar with an attack on an Israeli embassy in Baku, Cairo, or Amman but was killed on February 12, 2008, in Damascus before this plan could be carried out.[404] These three concerns reveal that the Israeli attack did risk the occurrence of a pernicious action-reaction cycle of violence between the state of Israel and Hezbollah, Syria, or Iran, a concern even voiced by the normally pro-Israel *Jerusalem Post*, which described the air attack as "the sort of thing that starts wars."[405]

Lessons for Future Management of Ambiguity, Deception, and/or Surprise   One lesson from the Israeli attack is that the "two wrongs make a right" principle still operates in international relations regarding the use of strategic surprise. Because Syria was violating international law (the Nuclear Nonproliferation Treaty) by building the nuclear plant, Israel's violation of international law (the breach of Syrian airspace, destroying property and ignoring the inviolability of national territorial sovereignty) went largely unheeded. Tit-for-tat disregard for restraining global norms can easily undermine a stable world order, however justified by special circumstances, and further global anarchy.

Another lesson is that sometimes interpretive confusion in part due to information overload can well serve a foreign manipulation initiator. Israel (and the United States) really benefited from all of the conflicting international news reports on the incident, because the prevention of global consensus no doubt was conducive to the absence of any concerted global condemnation of Israeli action. Particularly when the form of foreign manipulation is strategic surprise, such long-term interpretive confusion can enhance the willingness of a strategic initiator to undertake similar strategic surprises in the future.

Finally, as with Russian use of force in its geographical periphery (following Thucydides's famous conclusion about Athens and Melos), the Israeli attack on the Syrian al-Kibar nuclear plant reinforces the idea that the "might makes right" principle still applies in today's world, at least when applied to neighbors. In the current global security environment, "this incident is a reminder that there is no substitute for military strength and the will to use it."[406] Brash coercive surprise undertaken by a military power can reap national security rewards and still evade meaningful punishment from the international community.

2005 Andijan Massacre in Uzbekistan

The violent crackdown in Andijan in the Ferghana Valley in eastern Uzbekistan took place on May 13, 2005, in what came to be globally known as the Andijan massacre. Initially protesters "seized arms from a military depot and attacked a local prison, freeing the inmates, including some who had been jailed on charges related to radical Islamic activity," and then "after battling local police during the early morning hours, militants took control of local government offices."[407] At the same time, over ten thousand unarmed protesters gathered in Andijan's Bobur Square in front of the regional administration building, "drawn by an expectation that President Islam Karimov would come to address the protest";[408] "many of the protesters called on Karimov to resign, voicing complaints with the government's economic and political policies."[409] In response, shortly after 5:00 p.m. that day, eyewitness accounts reported that "government armored personnel carriers opened up on the crowd with heavy weapons, firing at random while moving at high speed."[410] The overall death toll, which included many innocent bystanders, was estimated to be more than seven hundred,[411] with men indiscriminately buried fifty to each grave and women and children rumored to have been "dissolved in acid" to conceal their demise.[412] This incident was strategically important because Uzbekistan has significant energy resources and the largest population in Central Asia, and President Karimov has "the region's largest and best equipped army and security forces."[413]

The trigger for the uprising was the arrest of twenty-three philanthropic businessmen:

> Protest participants, including some who escaped to Kyrgyzstan, said the trigger for the unrest was the government's decision to arrest 23 local businessmen on suspicion of belonging to Akromiya, the supposed Islamic radical group. Local residents . . . adamantly denied that the detainees had anything to do with Islamic radical activity. The entrepreneurs were described as devout Muslims who had become popular in Andijan by carrying out a wide array of charitable activity, including programs to assist the most impoverished local residents. The government, which is keen to monopolize authority, acted against the entrepreneurs out of concern that their philanthropy could some day be translated into political power, local residents said.[414]

What resulted was the Uzbek government unleashing "a crackdown on civil society, the ferocity of which is unprecedented even in Uzbekistan's fourteen-year history of repression since it became independent from the Soviet Union."[415]

Description of Ambiguity, Deception, and/or Surprise   The principal use of strategic surprise in the Andijan massacre was the brutal crackdown by the Uzbek government in response to the largely peaceful protest against the ruling regime. The international community was a primary victim of "strategic surprise, that is, a surprise act or actions of a magnitude that catches interested governments so off balance that it and its consequences (i.e. the reactions to it) affect the outcome of a war in truly consequential ways."[416] As with the Brexit case, some outside analysts felt that it should have been anticipated in advance, because "the warning signs are there for all to see"—the turmoil could have been avoided "by a coherent policy that took developments in Central Asia seriously, showed heightened rather than diminishing attention to the area, and took seriously both American strategic interests and the cause of democratization."[417]

The United States was not only surprised but also directly negatively affected by this massacre, which led to American forces being thrown out of the country:

> In Central Asia in 2005 . . . despite warnings about Uzbekistan's fragility by American experts . . . Washington did not expect the uprising at Andijan which came to have a profound impact on its regional position by significantly lessening its ability to use Central Asian bases in the war on terrorism. Moreover,

when the crisis did come the State Department and the Pentagon, gripped by rivalry, could not come up with a coherent response to the uprising and subsequent massacre at Andijan or to Tashkent's subsequent policies. As a result of the incoherence of U.S. policy, of which the demand for an investigation of Andijan is only a small part, America was ousted from its bases there.[418]

On November 21, 2005, the United States closed its Uzbekistan airbase—opened in 2001 and used for Afghanistan operations—and the U.S. Army withdrew from the Karshi-Khanabad airport, "a shutdown ordered by Uzbek President Islam Karimov after the United States joined calls for an international inquiry into the authoritarian leader's handling of the Andijan uprising."[419]

The Andijan massacre also exhibited strategic deception, specifically by the national Uzbek government, aimed at domestic citizens and the international community. Human Rights Watch, a respected transnational human rights organization, describes this regime deception: "The Uzbek government is using widespread repression and abuse to manipulate the truth, so that it can depict the protest itself as violent—organized by 'terrorists' with a radical Islamic agenda and with the participation of mostly armed protestors—and suppress any evidence to the contrary, and shift the blame for the deaths of so many unarmed people."[420] State "blocking of all communications during the May events, a continuing government propaganda effort, and show trials" resulting in guilty verdicts all were designed "to exonerate the regime, and in this they succeeded."[421]

Role of Information Overload in Triggering or Changing Ambiguity, Deception, and/or Surprise   The initial information overload problem surrounding the Andijan massacre was too much low-quality and/or contradictory information. Immediately following the violence, "the Uzbek government expelled all journalists and human rights campaigners from Andijan and forbade an international investigation."[422] This tight governmental control of the media and outside reporting ended up "complicating efforts to obtain accurate information about events,"[423] creating tremendous ambiguity surrounding what actually transpired during the massacre. At first even "few Uzbeks [knew] what really happened."[424] However, rather than problems emanating from too little information, this high uncertainty increased the ability of different parties to spin interpretations of the event in differing directions, overwhelming listeners with low-credibility information; for example, while Uzbek president Islam Karimov tried to deflect attention from his regime's repressive actions by claiming that "Islamic radicals are responsible for the recent violence," eyewitnesses contended that "government security forces indiscriminately shot civilians in acting to crush what was, in essence, a protest over the government's disregard for basic political and economic rights."[425] The ruling regime tried repeatedly to garner sympathy by creating "the spectre of foreign-linked religious extremists intent on taking over the country, rather than just desperate Uzbeks who wanted a better life."[426]

However, after this initial confusion, Internet-based social media undermined the Uzbek state's information control by bombarding the world with the true nature of the massacre:

> During the crackdown that followed the massacre, many of Uzbekistan's journalists, writers and activists were driven from the country. . . . Exiled from their homeland and isolated from one another, Uzbek dissidents banded together online, creating blogs and forums that served as safeguards of forbidden narratives of Andijan. Thanks to the Internet, exile ended up facilitating the very political interaction that it was supposed to prevent.[427]

This flood of bottom-up information discredited the top-down narrative of the Uzbek government, and "for Uzbeks around the world, a collective digital memory had been born."[428]

The Andijan massacre highlights that boasts about American information superiority in remote regions may be overinflated. Specifically, this case reveals that "U.S. claims about having a presumed information superiority or dominance in the theater are unwarranted because of America's demonstrated and consistent failures to understand critical aspects of these theaters and the nature of the wars it is fighting," in which "reliance upon the wonders of technology invites and almost mandates surprise attacks and the whole panoply of manifestations of so-called asymmetric warfare that confound American or even Western knowledge and understanding."[429] Indeed, "given the multiple problems afflicting the U.S. intelligence community, it is unlikely that even if it knew what was coming at Andijan and what could be the consequences of that kind of insurgency, that it would have either been able to adequately inform the government or that the government could have acted effectively on that information given its own divisions."[430] The Andijan massacre shows that American "ability to distinguish reality or real signals from 'noise' is still greatly impaired, notwithstanding the euphoria of enthusiasts of the supposedly transparent battlefield."[431]

Rationale and Purpose of Ambiguity, Deception, and/or Surprise   As is typical of domestic uprising responses, the Uzbekistan government regime's motive was to preserve the status quo and maintain order. President Karimov specifically feared that the protesters planned to "overturn the existing constitutional order" and "set up a so-called Muslim caliphate" under shari'ah Islamic law.[432] In retrospect, "it is quite unlikely that the Karimov regime that disregarded American advice and pressure for reform could have averted Andijan."[433]

The motivation of the Uzbek protesters revolved around deep discontent about their plight. Demonstrators told reporters that "government economic policies, especially punitive taxation on trade, were impoverishing many Uzbeks without providing any means for relief of dire financial problems."[434] Most broadly, the largely peaceful activists wished "to protest the deteriorating social, political and economic conditions of Uzbekistan."[435]

In U.S. foreign security policy in Central Asia, American motives have encompassed "ensuring the geopolitical independence and sovereignty of all the Central Asian regimes," especially "their freedom from terror and from Russian and Chinese neo-imperial pressures," global access to Central Asian energy reserves, democratization, and the development of pro-Western policies.[436] Among these objectives, "the primary strategic goal of the United States in Central Asia is to see the development of independent, democratic, and stable states, committed to the kind of political and economic reform that is essential to modern societies and on the path to integration and to the world economy."[437]

Policy Effectiveness of Ambiguity, Deception, and/or Surprise   In the wake of the Andijan protest, the Uzbek regime had short-run military success in minimizing the aftereffects of its violent and brutal crackdown on the largely peaceful protesters:

> Andijan's prison, rammed at high speed on May 13th by a hijacked truck to free prisoners, now has new gates and zig-zag concrete blocks outside. The town hall, where government hostages were held by protesters, is being reconstructed. Buildings facing Babur Square, where up to a thousand peaceful bystanders were gunned down by government troops as they listened to protesters' speeches, are still pockmarked by mortar and bullet holes. But five months after the atrocity— probably the worst committed by a government against demonstrators since Tiananmen in 1989—there are few other overt signs of the uprising against the repressive, and economically disastrous, regime of Islam Karimov.[438]

Even a decade after the massacre, Uzbekistan's politics "remained essentially unchanged":

> President Karimov—the first and only president of Uzbekistan—was re-elected in 2007 and 2015, with both elections ruled unfair by international observers. Repeated calls for an international investigation into Andijan have been ignored. Opponents of the regime have been persecuted both inside Uzbekistan and outside—such as the killing of the journalist Alisher Saipov in 2007 in Kyrgyzstan and the shooting of the imam Obidkhon Qori Nazarov in 2012 in Sweden. Both men were outspoken online critics of the state's role in Andijan.[439]

So the perpetrators of the slaughter of innocent people have so far seemed untouchable.

However, in the long run, citizen conditions have deteriorated since the Andijan massacre, with the biggest political stability threat being state failure rather than terrorist takeover:[440]

> But while their knowledge is mostly hazy, many Uzbeks say "everything" has gotten worse since Andijan. Surveillance and security have been ratcheted up still further. There are police every 50 metres in Tashkent and dozens of roadblocks on the single mountainous road that leads to the Fergana valley, where Andijan is located. New methods of control include civilian informers who report directly to the police. Indigenous neighbourhood committees now keep tabs on the population. Most foreign NGOs [non-governmental organizations] have been closed down, the remaining few threatened.[441]

In addressing such remaining enduring hardships, President Karimov has not been able to respond effectively and instead "has explicitly blamed the 'intrusive meddling' of America and the West."[442]

Perceived Legitimacy of Ambiguity, Deception, and/or Surprise   The Uzbek government's violent repression dramatically lowered its political legitimacy both at home and abroad. Although muzzled by the state, Uzbek citizens are not happy—five months after the uprising, "although the tree-lined streets of Andijan now appear calm, the mood is tense, angry and fearful."[443] In the end, "Islam Karimov's regime, thanks to its violence and repression, has lost any semblance of legitimacy they have abroad except in Moscow and Beijing and probably at home"; "this regime, whose corruption is pervasive, now stands revealed as relying almost exclusively on violence."[444] Notably, "despite the considerable body of evidence that contradicted Karimov's claims," the United States "initially accepted his arguments concerning the use of force in Andijan and elsewhere in the Ferghana Valley," and out of desire to maintain access to its airbase (before it was closed) it joined Russia in June 2005 in blocking a demand for an international probe of the Andijan massacre.[445] Since that time, however, the United States "hardened" its position,[446] and Uzbekistan "is now sanctioned by the European Union and its regime is 'radioactive' insofar as Washington is concerned."[447] Nonetheless, despite egregious human rights violations in the massacre, "due to poor policy and bad intelligence no effective response to it could be devised by Washington,"[448] and the lack of Western pressure—along with external support from Russia and China—seem likely to cause the illegitimate repression to persist.

Unintended Consequences of Ambiguity, Deception, and/or Surprise   One unexpected security consequence of the Andijan massacre was regional contagion of violence. The bloodshed produced "a wave of would-be refugees streaming toward the nearby border with Kyrgyzstan."[449] In his February 2006 testimony to the U.S. Senate, John Negroponte, then American director of national intelligence, "warned that several Central Asian states could undergo upheavals that would leave them resembling the failed state condition of Somalia."[450] Tight interconnections among Central Asian states have enhanced these instability contagion possibilities:

> Thanks to decades of Soviet policies and post-Soviet support for regional integration, Central Asian countries are strongly interlinked, so Uzbekistan's neighbors are vulnerable to any instability next door. Weak and struggling, Kyrgyzstan and Tajikistan are essentially dependent on Uzbekistan for energy and transport. Even relatively prosperous Kazakhstan could be seriously troubled if violence were to drive Uzbeks across its border.[451]

Since the Andijan massacre, "Uzbekistan has already become an exporter of instability to its neighbors"; three Central Asian states—Turkmenistan, Uzbekistan, and Kyrgyzstan—"appear to be increasingly unstable," and Tajikistan "has regressed still further from democratization and far too many of its people and too much of its economy depend on the drug trade for sustenance."[452]

Another not-directly-intended security consequence of the massacre was a significant reduction in the American presence in the entire Central Asian region. The expulsion of the United States from its Uzbekistan base caused American secretary of state Condoleezza Rice to state that the United States seeks no new bases in Central Asia.[453] American influence in Uzbekistan has decidedly diminished following the violence.

A corresponding unintended security consequence is that Uzbekistan was inadvertently pushed further into Russia's sphere of influence. Postmassacre developments—"the expulsion of the U.S. and NATO forces from their bases, attacks on Kyrgyzstan that were clearly at Moscow's behest, and attacks that the West fomented the Andijan uprising, the acceptance of one-sided gas deals with Moscow and Uzbekistan's increasing integration into the Russian-led Collective Security Treaty Organization (CSTO)—all indicate that Uzbekistan no longer can uphold a truly independent foreign policy line in its dealings with its neighbors and other states."[454] Indeed, "both Beijing and Moscow expressed support for Karimov after the Andijan uprising."[455]

Lessons for Future Management of Ambiguity, Deception, and/or Surprise   A key lesson from the Andijan massacre, thwarting American interests there, is that even the most powerful state in the world with a vast intelligence apparatus can be a persistent and relatively hapless victim of foreign strategic surprise:

> Thus it is clear that America has repeatedly been surprised by the enemy, a telling sign of cognitive and strategic failure. But as Andijan shows as well, America has also been surprised by events of importance affecting its partners and allies. And this surprise has unequivocally added enormous costs to the strategies it must now conduct to extricate itself from those past failures.[456]

Even with prescient warnings from multiple sources, the United States was taken completely off-guard by the Uzbek government's violent crackdown on the Andijan protesters.

A second lesson is that a fumbled response to strategic surprise can backfire, ending up pushing ensuing developments in the opposite direction from that desired. In this case, American ineptitude about how to respond to the Uzbek government crackdown on the Andijan protesters opened the door not only for closure of the American military base in Uzbekistan but also for pushing that country to have much closer ties to Russia. The magnitude of such security-degrading backfire effects emanating from poorly conceived and executed defensive responses to major strategic surprise cannot be overstated.

2001 Al-Qaeda Terrorist Attacks on the United States

Armed only with knives and box cutters, on September 11, 2001, nineteen terrorists destroyed four commercial aircraft, demolished both towers of the New

York World Trade Center, and damaged the Pentagon in Washington, D.C. (a fourth hijacked plane meant to fly into the White House crashed in Pennsylvania). With the cost of planning and executing the attacks being a mere $400,000 to $500,000,[457] the result was direct damage estimated at $18 billion, humiliating the most powerful state in the world. The specific death toll was over 2,600 people at the World Trade Center, 125 people at the Pentagon, and 256 people in the four destroyed planes. The ringleader of the covert operation was Muhammad Atta, the Egyptian-born terrorist who took over as pilot of American Airlines Flight 11, which flew into the North Tower of the World Trade Center.

The al-Qaeda terrorist group, a radical Islamic organization founded in 1988 dedicated to fighting a holy war against the United States, initiated the 9/11 attacks, "the most brazen and shocking terrorist attacks conducted by a substate group in history."[458] This transnational entity was composed of "transient groupings of individuals" operating "outside of traditional circles" but with "access to a worldwide network of training facilities and safehavens."[459] Osama bin Laden, a wealthy and well-educated Saudi Arabian exile, founded al-Qaeda, and Kuwaiti-born Khalid Sheikh Mohammed developed the 9/11 attack plan. Al-Qaeda carefully chose which targets within the United States to hit; the world viewed the World Trade Center in New York as "a symbol of US wealth, power, and dominion," and "along with the Pentagon, US Capitol Building, and White House, these sites had been chosen as the loci for a terrorist attack not only because of their centrality to the financial and political nerves of the United States, but because of their symbolic value."[460]

The United States launched a forceful and multipronged set of defensive responses to the 9/11 terrorist attacks. First, on September 20, 2001, President George W. Bush began the War on Terror by vowing to "direct every resource at our command—every means of diplomacy, every tool of intelligence, every instrument of law enforcement, every financial influence, and every necessary weapon of war—to the destruction and to the defeat of the global terror network."[461] He specifically mentioned that both terrorists and countries that provided safe havens for terrorists would be targeted. Second, on October 7, 2001, four weeks after the 9/11 terrorist attacks and in direct retaliation, Bush began the war on Afghanistan, initiating Operation Enduring Freedom. The explicit goal in Afghanistan was to defeat the Taliban and al-Qaeda forces, identified as the perpetrators of the violence. About sixty thousand American soldiers involved in the war effort at the peak of the conflict. A coalition of Afghan Northern Alliance fighters, air sorties by American planes, Western special operations forces and intelligence operatives, and a small contingent of Western ground forces participated in the action. Although most of the bombing had ended by late 2001, when the new government regime in Afghanistan was inaugurated, violence there has persisted long afterward. Last, in November 2002, as a direct result of the 9/11 terrorist attacks, the United States "formally institutionalized the link between domestic security and foreign travelers by creating the Department of Homeland Security (DHS), which merged twenty-two branches of the U.S. government, including immigration processing and enforcement bureaucracies; the creation of DHS was the first significant addition to the U.S. government since 1947, when President Harry Truman amalgamated the various branches of the U.S. armed forces into the Department of Defense to better coordinate the nation's defense against military threats."[462]

Description of Ambiguity, Deception, and/or Surprise   Through use of "highly innovative" strategy and tactics,[463] al-Qaeda executed strategic surprise in the 9/11 attacks, despite the availability of prior warning signs:

> The terror attacks on the World Trade Center and the Pentagon (and the hijackings that enabled them) clearly caught the bulk of the U.S. leadership, the American people, and many supporters of liberal democracy around the world by surprise. Yet the severity of this shock should not be allowed to obscure the uncomfortable fact that the terrorists' motives and modus operandi were well known to many experts on terrorism within and outside of the U.S. government. The World Trade Center had been attacked before, there had been many hijackings of large passenger aircraft around the globe during the last three decades, and Middle Eastern terrorists had regularly made use of vehicle-based suicide attacks such as truck- and boat-bombs.[464]

Indeed, a 1995 National Intelligence Estimate had predicted future terrorist attacks against the United States taking place within its borders;[465] "al Qaeda's enmity toward the United States and its desire to carry out an attack on U.S. soil was well known prior to 9/11";[466] and the CIA and the Federal Bureau of Investigation (FBI) had even received prior warnings that al-Qaeda desired to attack New York and Washington specifically using airplanes.[467]

Al-Qaeda's use of strategic surprise resulted from its use of strategic deception. Both al-Qaeda's organization and operation were "marked by deception, denial, [and] stealth":[468]

Deception, based on a detailed knowledge of the methods and mindsets of its adversaries, was thus a key counterintelligence strategy employed by Al Qaeda in its preparations for the 9/11 attacks. In fact, this deception proved so effective precisely because it was based on the hijackers' keen knowledge of their adversaries' own perceptions and preconceived ideas. From the utilization of operatives likely to arouse the least suspicion, to the methods used by the hijackers to enter and operate within the United States, the effectiveness of Al Qaeda's deception and denial stemmed directly from their plausibility.[469]

For example, al-Qaeda terrorists flew first class in order to receive less scrutiny—"although some held legitimate passports and visas, their intentions were obviously concealed."[470] Their training manuals had useful details on terrorist use of deceptive tactics in the West.[471]

Role of Information Overload in Triggering or Changing Ambiguity, Deception, and/or Surprise   As with the Japanese attack on Pearl Harbor, a poor signal-to-noise ratio was the key culprit in the terrorists' ability to employ deception and surprise successfully: "The difficulty of sorting through the 'noise'—the sea of incorrect information and false warnings—may have dulled response to the crucial signals that were available before the attacks."[472] Specifically, "there was considerable 'noise' prior to the September 11 attacks that allowed al-Qaʻeda to achieve strategic surprise—through increased communications among terrorist cells, the intelligence community knew that an attack was imminent, yet did not know where or how the terrorists would strike."[473] Demonstrating the direct debilitating impact of information overload, one American government official commented after the 9/11 attacks that a prior "warning about the possibility of a suicide hijacking would have been just one more speculative theory among many, hard to spot since the volume of warnings of 'al-Qaeda threats and other terrorist threats was in the tens of thousands—probably hundreds of thousands.'"[474] A special advantage of al-Qaeda as the initiator was that "the warning problem of terrorism is exceptionally difficult, compounding the challenges that individual analysts, bureaucracies, and policy makers face with regard to strategic surprise."[475]

Given the prevalence of information overload and uncertain interpretation of incoming communication, terrorists like al-Qaeda have become quite sophisticated at foreign manipulation:

In this era of asymmetric warfare, Middle East terrorist organizations have become increasingly adept at manipulating information about their operational plans and actions to exploit the vulnerabilities of militarily superior opponents. The surprise attack by al-Qa'ida against the United States on September 11, 2001 is the quintessential case in point. By adopting counterintelligence methods, including denial and deception, al-Qa'ida was able to camouflage its operational units inside the United States.[476]

Al-Qaeda has become particularly clever in transforming itself to use the Internet to bombard the world with propagandistic information:

> Al-Qaeda today is no longer best conceived of as an organization, a network, or even a network-of-networks. Rather, by leveraging new information and communication technologies, al-Qaeda has transformed itself into an organic social movement, making its virulent ideology accessible to anyone with a computer. . . . By bypassing other more conventional mediums, the Internet creates not just the tools, but an entirely new forum for fostering global awareness of issues unconstrained by government censorship or traditional cultural norms.[477]

Al-Qaeda's use of the Internet, "consistent with the broader pattern of grass-roots activism occurring around the world," incorporated efforts to "coordinate movement activities, events, and actions, discuss topics of interest and news with movement participants; disseminate propaganda, educational, and training materials; identify, recruit, and socialize new membership, and find and exploit information about their opposition."[478] After the United States and its allies ousted the Taliban in Afghanistan, senior al-Qaeda leaders "used the Internet to replace their dismantled training camps, reconnect their weakened organization, and reconstitute their leadership."[479] Over time, "*jihadi* web users have become increasingly aware of attempts by governments to monitor their behavior," and thus on the web "*jihadis* have recently posted protocol about safe ways to use technology" without risk of being identified.[480] Without rapid information technology advances, the impact of transnational terrorist groups like al-Qaeda would be much more containable.

The United States posed a particularly attractive and vulnerable target for manipulation by al-Qaeda because of prevailing American defensive misconceptions at the time. American security policy makers indulged in "overvaluation of past successes in reducing airline hijackings, overconfidence in the current air security system, and insensitivity to previous warnings questioning existing airline security policy."[481] Embedded in this defective security analysis was "a kind of wishful thinking that helps to relieve anxiety, but at the cost of increasing vulnerability."[482] Even former American national security advisor

Condoleezza Rice fretted about the debilitating impact of bureaucratic inertia potentially reducing the ability to nimbly perceive and respond to unexpected disruptions in the prelude to the 9/11 terrorist attacks on the United States.[483]

The American government had been complacent about the al-Qaeda threat. The *9/11 Commission Report* stated that because "before 9/11, al Qaeda and its affiliates had killed fewer than 50 Americans, including the East Africa embassy bombings and the *Cole* attack [the bombing of the American Navy destroyer USS *Cole* in October 2000], the U.S. government took the threat seriously, but not in the sense of mustering anything like the kind of effort that would be gathered to confront an enemy of the first, second, or even third rank."[484] More broadly, "a decade before 9/11 the worldwide surge in Islamic fundamentalism and its virulent hatred of the West was largely unrecognized in America."[485] Notably, "lingering (and deceptively comforting) beliefs that terrorism was something that occurred abroad also contributed to lack of preparation to thwart the threat of catastrophic terrorism to the U.S. homeland."[486] In the 1990s, the FBI assured government terrorism experts that it "had the matter of al Qaeda cells in America 'covered.'"[487] Although both Bill Clinton and George W. Bush understood that al-Qaeda was a threat to the United States, the *9/11 Commission Report* concluded that pre-9/11, "we do not believe they fully understood just how many people al Qaeda might kill, and how soon it might do it"—"there was uncertainty among senior officials about whether this was just a new and especially venomous version of the ordinary terrorist threat America had lived with for decades, or was radically new, posing a threat beyond any yet experienced."[488]

This sense of complacency was evident not just among American government officials but also among American media outlets. When earlier in 2001 the issue of a future terrorist attack on the United States was raised, "one senior reporter from a well-known publication" said, "This isn't important—none of this is ever going to happen."[489] The *9/11 Commission Report* noted that "neither in 2000 nor in the first eight months of 2001 did any polling organization in the United States think the subject of terrorism sufficiently on the minds of the public to ask a question about it on a major national survey"; al-Qaeda and terrorism were not discussed much in the 2000 presidential campaign; and "Congress and the media called little attention" to the topic.[490] So both bottom-up and top-down roots explain the underprioritization of the terrorist threat.

Rationale and Purpose of Ambiguity, Deception, and/or Surprise   In pursuing its overarching goal of holy war against the United States, the 9/11 terrorist attacks on the United States furthered al-Qaeda's purposes "to propel the organization onto the world stage, drawing global attention to its grievances," and "to radicalize and mobilize supporters by offering an example of a successful operation achieved through methodical planning, personal sacrifice, and submission to the will of God."[491] Al-Qaeda was a particularly difficult initiator to predict and thwart in terms of its objectives and strategies: "too often, assessments of failure focus on the mistakes of the victim rather than on the skill of the adversary"; in this case, al-Qaeda proved "hard to disrupt because of its transnational nature and large size, both of which enabled it to lose a cell or skilled operatives yet continue operations elsewhere"; and because the group "has demonstrated an ability to revise its methods and structure in response to setbacks or failures."[492]

The United States was motivated by anger and a desire for revenge in its response to the 9/11 terrorist attacks. American citizens responded to the 9/11 attacks with uncertainty and fear,[493] and a mix of shock, pride, and rage: they hung American flags everywhere, displayed solidarity and patriotism wherever they could, attended large prayer vigils, and listened approvingly to Toby Keith's country song "Angry American," promising swift and decisive retaliation.

Policy Effectiveness of Ambiguity, Deception, and/or Surprise   The 9/11 attacks on the United States were resoundingly effective in the short term in promoting al-Qaeda's goals of globally publicizing its cause, attracting new recruits, and reducing the mass public's trust in the government's ability to protect them. Psychologically, "the attacks shattered U.S. complacency and replaced it with fears of a new and menacing organization that seemed to threaten our very survival."[494] The implications of the public relations victory among anti-Western populations around the globe accomplished by al-Qaeda through the 9/11 attacks on the United States were immeasurable:

> These carefully constructed and choreographed attacks, with their tremendous symbolic significance, revealed to multiple audiences Al Qaeda's worldview centered on the existence of a struggle between the forces of good and evil. Moreover, the attacks were planned to convey the power and righteousness of Al Qaeda's cause, while demonstrating the vulnerability of the U.S. government and all that it represents. By discrediting U.S. power in this way, Al Qaeda sought to establish the supremacy of its ideology in what it considers to be a world at war.[495]

In light of the 9/11 terrorist attacks on the United States, "al Qaeda's successful use of intelligence and counterintelligence represented an impressive display of power in its own right," exemplifying "the capacity of the organization to challenge the authority of the United States by overcoming its intelligence services, one of the most recognizable symbols of American power."[496] Overall, the deceptive and surprising 9/11 terrorist attacks provided a gold standard for other terrorist groups to try to emulate.

For American citizens, through the 9/11 terrorist attacks al-Qaeda succeeded in decisively altering their threat perception and making them feel vulnerable and helpless, as "the collective safety felt by American air travelers evaporated."[497] While American citizens responded to the 9/11 attacks with great patriotism and support for their country, at the same time they were upset at their government for allowing an unthinkable foreign attack on American soil to occur. The American mass public was specifically frustrated that the U.S. intelligence community had not been able to predict and forestall the 9/11 terrorist attacks.

However, in the long run, the coercive multipronged military response to the 9/11 terrorist attacks did make life more difficult for al-Qaeda. Many of its top leaders were killed through American drone strikes, and on May 1, 2011, Osama bin Laden was killed in an American Special Forces raid on his compound in Pakistan.

Nonetheless, thanks in large part to the 9/11 attacks' legacy, the global terrorist threat persists today, and global vulnerability to terrorist attack remains very high:

> Terrorism is a decentralized phenomenon—in its funding, planning, and execution. Removing bin Laden does not end the threat. There are successors in al-Qaeda—and successors in autonomous groups operating out of Yemen, Somalia, and other countries. So terrorism will continue. Indeed, it could even grow somewhat worse in the short run as there are sure to be those who will want to show that they can still strike against the West.[498]

Al-Qaeda's threat has been somewhat eclipsed by the terrorist organization ISIS (Islamic State of Iraq and Syria). ISIS—which emerged in April 2013 and used to be an al-Qaeda affiliate (linked to the Iraq faction) until al-Qaeda severed ties with it[499]—developed "much more financial power than Al Qaeda ever had"[500] and has been far more feared as a global threat than is al-Qaeda. Part of the ISIS success formula has been that "unlike al Qaeda, which has generally been methodical about organizing and controlling its terror cells, the more opportunistic Islamic State [ISIS] is content to crowdsource its social media activity—and its violence—out to individuals with whom it has no concrete ties," allowing ISIS "to rouse followers that al Qaeda never was able to reach."[501] So in retrospect it appears that while the glow of al-Qaeda's 9/11 success faded a bit over time, its inspiration of global terror attacks has persisted.

Perceived Legitimacy of Ambiguity, Deception, and/or Surprise   Western countries universally viewed al-Qaeda's 9/11 attacks on the United States as heinous, unjustified, and immoral. However, some anti-Western foreigners cheered that the superpower's vulnerability was finally exposed. While most Mideast political leaders refrained from open criticism, the Iraqi government announced that "the American cowboys are reaping the fruit of their crimes against humanity."[502] For those people globally with "affinity for *jihad* and al-Qaeda," an active online community served to legitimize the 9/11 attacks, "helped rally support for Osama bin Laden, and facilitated . . . communication among non-Arab *jihad* sympathizers."[503] Even the *9/11 Commission Report* admitted that al-Qaeda had garnered "broad support in the Arab and Muslim world by demanding redress of political grievances."[504]

Concerns by the American people about their government being taken by surprise by such a devastating foreign attack "prompted outraged demands—as did the sneak attack on Pearl Harbor almost sixty years earlier—that an independent inquiry investigate the intelligence and government breakdowns that preceded the attack."[505] Nonetheless, when Osama bin Laden was finally killed nearly ten years later on May 1, 2011, a crowd of around two hundred people assembled outside the White House waving American flags and singing "The Star Spangled Banner" and chanting "U-S-A";[506] another large crowd gathered at Ground Zero in New York and sang "God Bless America."[507] Indeed, "more than the toppling of Saddam Hussein's statue in Baghdad, the purple-stained fingers of Iraqis at ballot boxes, or the smiling faces of Afghan girls at school ribbon-cutting ceremonies—the raid that killed Osama bin Laden stands, to date, as the defining moral victory of America's war on terror."[508] In contrast, after Osama bin Laden died, much of the Arab world responded with sadness and anger: "In sharp contrast to the celebrations in America, on the streets of Saudi Arabia, Bin Laden's native land, there was a mood of disbelief and sorrow among many. The Palestinian Islamist group Hamas mourned Bin Laden as an 'Arab holy warrior'";[509] and "Arab readers' responses from across the Middle East to news of bin Laden's death on mainstream websites such as *Al-Jazeera* have been overwhelmingly condemnatory of America and full of praise for

Osama bin Laden, the 'martyr,' the 'warrior,' the 'hero.'"[510] These diametrically opposed reactions to bin Laden's death illustrate the extremely different notions of legitimacy at work globally across cultures.

Unintended Consequences of Ambiguity, Deception, and/or Surprise   One unintended consequence of the 9/11 terrorist attacks was the growing tension between the United States and its Middle East allies from which the al-Qaeda terrorists originated. Americans blamed some previously friendly Islamic regimes as tacitly aiding al-Qaeda. For example, the *9/11 Commission Report* accused donors in the Persian Gulf states, especially Saudi Arabia, for providing core financing for al-Qaeda;[511] later the United States expressed disappointment with Pakistan for harboring Osama bin Laden before he was killed.

A second closely related unintended consequence of the 9/11 attacks was the blossoming among Americans of Islamophobia, which later escalated still further within the United States and the West after ISIS engaged in numerous acts of global violence. Many Americans cannot seem to distinguish clearly among the categories of devout Muslim believers, radical Muslims, extremist Muslim zealots, and Islamic fundamentalists, and as a result lump them all together as posing severe threats to the American way of life. This xenophobic fear and hatred of perceived "Islamic fanatics" persists even today and hurts security relations between the United States and predominantly Muslim countries.

Lessons for Future Management of Ambiguity, Deception, and/or Surprise   One key lesson from the 9/11 terrorist attacks is that if strategic surprise is adeptly executed and key signals are enmeshed in tons of noise, then advance warning and effective prevention may be highly improbable. Given the careful planning of the 9/11 terrorist attacks and the maze of disconnected clues requiring processing and integration under time pressure, it seems unrealistic to conclude—as some analysts have in a manner comparable to such critiques in the Brexit and Uzbekistan cases—that "the responsible officials should have been able to put the puzzle together and thwart the attacks" before they occurred.[512]

A second important lesson is that adept nonstate groups may often be able to adapt more quickly and flexibly to changing information technologies than can status quo major power states. As former secretary of defense Donald Rumsfeld astutely noted, "Our enemies have skillfully adapted to fighting wars in today's media age, but . . . our country has not."[513] As with ISIS in years afterward, "al-Qaeda's use of the Internet and other new technologies has also enabled it to radicalize and empower armies of new recruits by shaping their general worldview."[514]

A third strategic lesson is that massive defense spending does not automatically translate into proper alertness, warning, vigilance, and preparation for foreign terrorist deception and surprise. Prior to 9/11, although the United States was concerned about the terrorist threat and devoted substantial resources to counterterrorism, complacency about the adequacy of its preparation—along with an inability to centrally connect disparate warning signals within a very noisy communication setting—interfered with proper interpretation of incoming information and timely formulation of effective responses. In some senses, patting oneself on the back for engaging in substantial counterterrorist efforts opens the door to unexpected foreign terror attacks.

1990 Iraqi Attack on Kuwait

To many observers at the time, the 1990 Iraqi invasion of Kuwait and the international response to it signified the beginning of a new era. Although some analysts might contest the inclusion of a 1990 case as being part of the Internet age, its inclusion in this study is justified by the widespread availability by the early 1990s to both government officials and private citizens in Global North countries of BBSs (cross-location digital bulletin board services) accessible through dial-up modems allowing access to unprecedented treasure troves of data.

At 2:00 a.m. on August 2, 1990, Iraq began a carefully orchestrated three-day invasion of Kuwait. As the fourth-largest military in the world,[515] Iraq was able to quickly dispatch 120,000 troops, 850 tanks, and numerous helicopters and trucks into Kuwait. Within an hour, the Iraqi troops had reached Kuwait City. By daybreak, Iraqi tanks assaulted the royal residence, Dasman Palace: "The emir had already fled into the Saudi desert, but his private guard and his younger half-brother, Sheik Faud al-Ahmad al-Sabah, had stayed behind to defend their home; the sheik was shot and killed, and according to an Iraqi soldier who deserted after the assault, his body was placed in front of a tank and run over."[516] Soon the invading army controlled Kuwait, with most of Kuwait's armed forces overcome by Iraqi troops or in flight to the neighboring countries of Saudi Arabia and Bahrain. Iraq then proceeded to annex Kuwait as its nineteenth province.

After Kuwait appealed for international aid, the UN Security Council passed a resolution (14–0, with Yemen abstaining) demanding that Iraq immediately withdraw from Kuwait, and Iraq continued to defy UN sanctions and demands, the international community's response was "unprecedented": the Gulf War commenced with UN approval, involving an American-led twenty-eight-state international coalition composed of seven hundred thousand air, land, and naval forces.[517] Operation Desert Storm, a forty-three-day operation that took place between January 16, 1991, and February 28, 1991, was "the most extensive and rapid US military mobilization since the end of World War II."[518] After a full-scale coalition air-and-missile attack targeted strategic Iraqi military targets such as airports, command-and-control centers, missile launch sites, and radar stations, the primary ground thrust occurred after a month had elapsed during the last 100 hours of the war. By the end of February, the conflict was over, with these coalition forces having defeated the Iraqi military and freed Kuwait from its seven-months-long occupation. On March 15, 1991, the emir of Kuwait returned from exile to the country. On April 11, 1991, the United Nations declared a formal end to the war.

The quick U.S. victory in the war was due to its technological superiority and Iraqi leadership deficiencies. The United States had a vast technological advantage over Iraq,[519] especially regarding advanced air power capabilities: "The Gulf War showed air power off to great advantage but in extremely favorable circumstances—the United States brought to bear a force sized and trained to fight with the Soviet Union in a global war, obtained the backing of almost every major military and financial power, and chose the time and place at which combat would begin in a theater ideally suited to air operations."[520] Because Iraq manifested key deficiencies in leadership and soldier morale and training, to many observers "the enemy in the Gulf War was indeed 'perfect,'" as "it was mainly a deeply confused and remarkably ill-led rabble."[521]

Description of Ambiguity, Deception, and/or Surprise   Even though the United States considered Iraq a primary focus of its security attention, it was a victim of strategic surprise when it came to the August 1990 Iraqi invasion:

> Despite Iraq's heavy buildup of forces along the Kuwaiti border in late July, the Bush Administration was surprised by the invasion . . . and largely unprepared to respond quickly, Administration officials say. The failure to anticipate the Iraqi attack has provoked heated debate within the Administration over the quality of intelligence assessments in the days before the invasion, including an unusual public defense of its work by the Central Intelligence Agency. But while much evidence suggests that the invasion on Wednesday evening [August 1, 1990] caught senior national security officials unaware, it is less clear whether

that was because of faulty intelligence or a misreading of the situation by policy makers who saw Iraq's buildup more as a show of force than as final preparations for battle.[522]

Iraq's attack was also a shock to Kuwait and other Arab states—"few Arab governments actually expected Iraq to invade Kuwait," for they thought Iraq's military buildup was just a show of force and that the Arab League's prohibition on one Arab state invading another would prevent war.[523]

Role of Information Overload in Triggering or Changing Ambiguity, Deception, and/or Surprise   Information overload, differences in modes of cultural expression, and misinterpretation of signals were prominently evident in the Iraqi invasion of Kuwait. On July 25, 1990, in a meeting between Saddam Hussein and American ambassador to Iraq April Glaspie, she stated to him that "the U.S. did not consider border disputes its business" but that "it was emphatically our business that they [Iraq and Kuwait] reach a settlement in a non-violent way" through diplomacy; shortly afterward, however, Iraq released a transcript of the meeting quoting Ambassador Glaspie as saying, "The U.S. has no opinion on Arab-Arab conflicts, like your border disagreement with Kuwait," appearing to give Iraq a "green light" to undertake military action against Kuwait.[524] When interviewed on January 17, 2004, after his capture following the 2003 Iraq War, Hussein said he remembered the invasion as "fundamentally a dispute between Iraq and Kuwait," expecting that "the United States would not interfere in a dispute between two Arab countries";[525] yet President George H. W. Bush later wrote in his memoir, "No one, especially Saddam Hussein, could doubt that the U.S. had strong interests in the Gulf and did not condone aggression."[526] Rather than a discrepancy in memories, this contrast seems more likely to reflect vast cultural signaling differences.

Saddam Hussein appeared to be an overconfident "high-stakes gambler" who used selective attention to interpret trends favorably:

> "Shrewd . . . but stupid" is how he was once aptly described. In 1980, Saddam underestimated the ability of Iran to repel a military invasion, leading to an eight-year war that killed 100,000 Iraqis and cost Baghdad nearly half a trillion dollars. He underestimated U.S. resolve to liberate Kuwait, despite a 40-year U.S. commitment to protect the gulf monarchies.[527]

Because Hussein failed to read the security tea leaves accurately, he "miscalculated drastically" in his "claims of invulnerability," "his belief in Iraq's centrality to pan-Arab security and identity," "his aspirations for economic power," and his "thought that he could quickly absorb Kuwait and present a *fait accompli* to the world."[528]

Warnings emerged that could have prepared Kuwait and the United States for the Iraqi invasion, but they fell on deaf ears. A July 25 CIA "warning of war" memorandum stressing the imminence of the Kuwait invasion garnered little attention from top policy makers.[529] As early as mid-July, many intelligence reports were issued about the Iraqi military buildup, but government analysts concluded that "Iraq was bluffing to force Kuwait and other members of the OPEC oil cartel to raise prices and reduce surplus oil production."[530] Believing that "Iraq's saber-rattling was bluster, not genuine," senior American policy makers remained "unimpressed with the seriousness of the developing crisis" until "only hours before it occurred, far too late for useful military or diplomatic deterrents."[531] Thus significant misperception of the security predicament existed within both Iraq and the United States.[532]

Information overload directly contributed to American policy makers' reluctance to believe the warning signals. Because of Saddam Hussein's past histrionics, the "cry wolf" distortion made recipients downplay these warnings in favor of less alarming conclusions. Conflicting information was present, with a poor signal-to-noise ratio; for example, the American ambassador to Iraq had sent a cable noting Saddam's assurance that he had no intention of taking military action because of forthcoming talks between Iraq and Kuwait in Jeddah and Baghdad."[533] Even among those predicting an Iraqi invasion, ambiguity was present because of great uncertainty about the size and scope of any attack.[534] Finally, distraction was present, because it turns out that the American government and much of the rest of the world had been otherwise "lulled by events outside the Mideast into a false sense of peace," specifically "preoccupied with events in Europe."[535]

Rationale and Purpose of Ambiguity, Deception, and/or Surprise   The Iraqi government's rationale for invading Kuwait was multifaceted. Shortly before the invasion, "Iraq had accused Kuwait of flooding the world market with oil and has demanded compensation for oil produced from a disputed oil field on the border of the two countries."[536] As the biggest oil producer in the Middle East, Kuwait had in 1989 exceeded its OPEC oil quota by seven hundred thousand barrels, hoping to "force Mr. Hussein to the bargaining table and then extract from him a border truce that included Rumaila [the disputed oilfield] drilling rights as well as a non-aggression pact."[537] The oil glut led to weak global prices and falling revenues, triggering "an acute financial crisis in Baghdad"—Iraqi foreign minister Tariq Aziz claimed that "every $1 drop in the price of a barrel of oil . . . caused a $1 billion drop in Iraq's annual revenues."[538] Territorial disputes were also a trigger: Iraq had never accepted Kuwaiti control of two uninhabited islands off Kuwait, Bubiyan and Warba, restricting Iraqi access to the Persian Gulf.[539] Moreover, despite Kuwaiti royal family rule there since 1756, Iraq always considered Kuwait as part of its southern province, even when Kuwait became a British protectorate in 1899 and became independent in 1961.[540] Most broadly, Iraq had regional hegemony aspirations: "Saddam views the Persian Gulf as Iraq's 'natural' sphere of influence" and had "a clear intent to play a role in the wider Arab world" in the context of the "rivalry among Arab states for dominance" in the Persian Gulf.[541] After expressing anger at Kuwait in a May 1990 speech to the Arab League, in which he said, "We have reached a state of affairs where we cannot take the pressure,"[542] and in a July 17, 1990, speech in which he denounced "certain rulers of the Gulf states" who have "thrust their poisoned dagger in our back,"[543] Saddam Hussein alone made the decision to attack Kuwait, gambling "that he could get away with seizing the tiny oil-rich nation to help pay off debts,"[544] partially because of the long war with Iran.

The American motivation in responding with military force was to liberate Kuwait, at least on the surface a classic balance-of-power attempt to maintain the status quo; at stake was the maintenance of "the existing political and economic order in the region, which the US is striving to maintain and which Iraq appears to be challenging."[545] The long-term goal was stated as a desire to deter aggression and to preserve the principle of the self-determination of peoples.[546] However, many outside analysts perceived that American strategic interest in access to Kuwaiti oil and its distribution played a significant role in the American reaction to the Iraqi invasion.[547] Moreover, "the scale and timing of the US military intervention have submerged the important issue of Iraqi aggression and Kuwaiti sovereignty within the larger issue of Arab self-determination against Washington's efforts to impose its hegemony over the Gulf region and the Arab world."[548]

Policy Effectiveness of Ambiguity, Deception, and/or Surprise   Iraq's use of strategic military surprise in its invasion of Kuwait manifested short-run military success but not long-run policy effectiveness. Although Iraq was temporarily able to achieve its strategic objectives and to physically take over and occupy Kuwait, that outcome was quickly reversed by the American-led 1991 Gulf War. This reversal occurred partially because "the Iraqi strategy to keep the crisis 'an Arab affair' totally failed."[549] Overall, despite his belief that Gulf War was in some ways "beneficial,"[550] Hussein's "gamble did not pay off—he had misread the interests of the international community and the United States in a stable Middle East."[551]

In an eerily similar manner, the American-led coalition's military victory in the 1991 Gulf War also had only short-term positive payoffs. Although the defeat did eject Iraq from Kuwait, Saddam Hussein remained such a troublesome sore spot in the Middle East that the United States was forced to take military action again in the 2003 Iraq War: "the fact that Saddam Hussein remained in power in Iraq and was able to crush his internal opponents with such ferocity left many feeling that the operation to free Kuwait had been only a partial success."[552] So in the end neither Iraq nor the United States benefited much from their violent confrontation over Kuwait.

Perceived Legitimacy of Ambiguity, Deception, and/or Surprise   The world perceived the Iraqi invasion of Kuwait as illegitimate, violating the prohibition against using force unilaterally within another sovereign state.[553] Within the Middle East, "for most Arabs, the initial reaction was shock, and disapproval of the invasion."[554] Global condemnation quickly ensued. In an emergency session, the UN Security Council called for the "immediate and unconditional" withdrawal of Iraqi forces from Kuwait and on August 6, 1990, instituted an international trade ban with Iraq; British prime minister Margaret Thatcher "branded the invasion as 'absolutely unacceptable,' and American president George H. W. Bush "condemned the attack as 'a naked act of aggression.'"[555] Many countries froze Kuwaiti assets abroad to deny Iraq the fruits of military victory. Even the Soviet Union, Iraq's primary arms supplier, responded to the Iraqi invasion by suspending the delivery of all military equipment to Iraq.

Making matters worse, during its control of Kuwait the Iraqi regime committed numerous human rights violations, brutalizing suspected members of Kuwait's armed resistance movement. The Iraqi military "tortured and killed hundreds of Kuwaiti nationals and people from other nations," took foreign hostages, looted Kuwaiti properties, and "set fire to more than 700 Kuwaiti oil wells and opened pipelines to let oil pour into the Gulf."[556] Such actions further eroded the legitimacy of the Iraqi invasion of Kuwait and incensed the international community.

Unintended Consequences of Ambiguity, Deception, and/or Surprise   Iraq's regional image and influence did not suffer a severe hit after its unlawful invasion of Kuwait and its subsequent defeat in the 1991 Gulf War. Until his capture after the 2003 Iraq War, Saddam Hussein had appeared relatively immune to such jolts, having "presided as the head of state longer than all his predecessors, despite three costly wars, numerous internal insurrections, coup attempts and crippling international sanctions":[557]

> Saddam saw himself as the latest in a long line of great Iraqi leaders that included Nebuchadnezzar, Hammurabi, and Saladin. He was narcissistic, and he had a very long time horizon. . . . Saddam valued the character of the struggle and endurance, even if it resulted in something others might term defeat. . . . Merely remaining in power in the face of an onslaught that Saddam saw as historically unprecedented constituted, to his calculation, a stunning victory.[558]

When given a chance, he had been consistently wily enough to stay in power.

Nonetheless, after the invasion, Saddam Hussein did seem a bit more hemmed in, backing down when facing a credible counterthreat:

> He didn't use chemical or biological weapons against U.S. forces during the Gulf War, apparently in response to a White House threat that he and his country would "pay a terrible price." He temporarily stopped interfering with U.N. weapons inspectors in 1993, seeking to avoid confrontation with the United States. And when he seemed poised to invade Kuwait once again in 1994, he backed down in response to a U.S. military buildup.[559]

After the Gulf War, the United States reinforced this restraint by retaining a strong defensive military presence in Kuwait. Moreover, Hussein's risky behavior often alienated his allies, most of whom had opposed his Kuwaiti invasion: for example, in 1998, "he isolated himself and angered Arab leaders when he called upon Arabs everywhere to rise up and topple the corrupt 'throne dwarves' who had chosen to collaborate with the United States."[560] Ironically, the United States reaped little increased respect among Arab states from its restoration of Kuwaiti sovereignty.

Despite the military success of the American response to the Iraqi invasion of Kuwait, there were costs in terms of the relations of the United States and its Middle East allies:

> Despite the war's success (and for some Arabs because of it), to the extent that America intervenes in the complex, historical problems of biblical proportions in the Middle East, we are going to increase resentment against the West thereby

fostering even more fundamentalist dictatorships than already exist. More trade wars will result. And these developments could lead to an even greater war in the Middle East. We might not have the majority of Arab nations on our side in that one. As Arab resentment increases, one can be sure that terrorism will also increase.[561]

In the decades since the end of the 1991 Gulf War, anarchic violence has increased in the Middle East. Moreover, many Arabs have been angry at "the transparent hypocrisy of the United States in its very selective opposition to aggression" in the region "to maintain a political and economic status quo which, for millions, has become intolerable."[562]

Lessons for Future Management of Ambiguity, Deception, and/or Surprise   One lesson from the American failure to anticipate the Iraqi invasion of Kuwait is that intelligence on intentions is more important—and usually more difficult to obtain—than intelligence on capabilities. It was easier for American intelligence analysts to observe the highly visible buildup of Iraqi forces on Kuwait's border than to distinguish between Iraqi intentions simply to provide a preemptive show of force versus actually launching a full-scale military invasion of Kuwait. Especially when attempting to respond to subtle foreign information manipulation, relying for signal interpretation primarily on tangible developments such as troop movements or increases in defense spending cannot substitute for obtaining high-quality culturally sensitive intelligence on what the adversary plans to do given such military moves.

Another lesson is that smaller states, even when being carefully monitored from the outside by major powers, can still engage in successful applications of strategic surprise and deception. It does not take many resources or much sophistication to create so much noise within the information environment that it becomes extremely difficult for outsiders to interpret correctly what is currently transpiring and what is likely to transpire in the near future. When an enemy leader like Saddam Hussein is known to be reckless, strategic deception and surprise can work even better because the range of options on the table is much more diverse, the value of prudent outside cost-benefit analysis of most likely actions can be misleading, and the ability to interpret the real meaning of brash threats can be lower.

A third lesson, parallel to that in the Georgia case study, is that successfully carrying out strategic manipulation does not necessarily guarantee long-term security. Based on skilled use of deception, Iraq managed to carry out a surprise attack on Kuwait and won a decisive military victory there, but soon afterward thanks to outside intervention the outcome was reversed and Iraqi military forces were decisively defeated. Even though Iraq's aggressive act was in its immediate neighborhood within its sphere of influence, and Iraq possessed considerable military power, the country's coercive capabilities were not sufficient to deter retaliation from a major power and its allies upset at the overturning of the status quo and the jeopardizing of access to a vital energy resource. Thus gains from strategic manipulation can sometimes prove to be fleeting.

4
Emerging Case Patterns

THE CASE STUDIES IN CHAPTER 3 flesh out key variations in how strategic manipulation affects global security under information overload. Tables 4.1, 4.2, 4.3, and 4.4 display twenty indicators about the case background context, case initiator offensive manipulation, case target defensive response, and case strategic outcome. The case background context includes the initiator identity, the initiator type, the target identity, the target type, and the incident date. The case offensive manipulation includes initiators' security goal, confrontation trigger, manipulation type, exposure to information overload, and action thrust. The case defensive response includes targets' response capacity, power disadvantage, security reaction, citizen attitude, and outside aid. Last, the case strategic outcome includes the security impact, the short-run effectiveness, the long-run success, the legitimacy image, and the unintended effects.

Broad Case Trends

Overall, information overload played a significant role in enhancing both the probability of manipulation and the range of available manipulation techniques. Strategic ambiguity, deception, and surprise were often extremely effective—with relatively low short-term costs—in advancing immediate security objectives. Although national governments possess certain advantages in access to and control of manipulation instruments, the two cases with nonstate manipulation initiators—the 2001 9/11 terrorist attacks and the 2016 Brexit vote to leave the EU—show that nonstate groups can be just as successful as states in distorting or falsifying information. However, despite its pervasiveness, the

Table 4.1 Case background context

Case	Initiator	Initiator type	Target	Target type	Incident date
9/11 terrorist attacks	Al-Qaeda	Nonstate group	United States	Sovereign territory	September 11, 2001
Andijan massacre	Uzbek government	State	Uzbek protesters International community	Nonstate groups	May 13, 2005
Annexation of Crimea	Russian government	State	Ukraine	Sovereign territory	February–March 2014
Attack on Kuwait	Iraqi government	State	Kuwait	Sovereign territory	August 1990
"Brexit" vote to leave the EU	UKIP Populist Party Print/digital and social media	Nonstate groups	British government	National government	June 23, 2016
Destruction of al-Kibar nuclear plant	Israeli government	State	Syria	Sovereign territory	September 6, 2007
Fukushima nuclear disaster	Japanese government	State	Japanese people International community	Private citizens Foreign governments	March 11, 2011
Invasion of Georgia	Russian government	State	Georgia	Sovereign territory	Summer 2008
Quest to find WMD	Iraqi government	State	United Nations	International organization	2002–2003
Trump's foreign security policy	American president Donald Trump	Head of state	U.S. allies and adversaries	National governments	January 2017–present

Table 4.2 Case initiator offensive manipulation

Case	Security goal	Confrontation trigger	Manipulation type	Information overload	Action thrust
9/11 terrorist attacks	Spread fear and distrust of state	Hatred of United States	Deception and surprise	Tons of noise and false signals	International violent force
Andijan massacre	Maintain order	Arrest of philanthropic businessmen	Deception and surprise	Noise and disinformation	Internal violent force
Annexation of Crimea	Maintain sphere of influence	Ousting of pro-Russian Ukrainian head of state	Deception and surprise	Noise/disinformation cloud Russian intent	International violent force
Attack on Kuwait	Expand territorial control	Kuwait flooding world market with oil	Surprise	Noise and conflicting data	International violent force
"Brexit" vote to leave the EU	Promote sovereignty and resist foreign meddling	UK economic woes and influx of migrants	Ambiguity, deception, and surprise	Barrage of conflicting news stories	Internal peaceful influence
Destruction of al-Kibar nuclear plant	Prevent nuclear weapons development	Construction of nuclear facility	Deception and surprise	Extensive global media coverage	International violent force
Fukushima nuclear disaster	Maintain order and prevent panic	Earthquake and tsunami	Deception	Inconsistency in state guidance	Internal peaceful influence
Invasion of Georgia	Prevent Georgia from cozying up to the West	NATO proposal about admitting Georgia	Deception and surprise	Noise/disinformation cloud Russian Intent	International violent force
Quest to find WMD	Hide scope of weaponry concealment	UN Inspection and U.S. WMD claims	Ambiguity and deception	Many discrepant assertions	Internal perceived concealment
Trump's foreign security policy	Project unpredictability and keep foes off-guard	2016 presidential election	Ambiguity and surprise	Tons of conflicting Trump tweets	International power projection

Table 4.3 Case target defensive response

Case	Response capacity	Power disadvantage	Security reaction	Citizen attitude	Outside aid
9/11 terrorist attacks	Very high	None	2001 war on Afghanistan	Anger and desire for revenge	Yes (requested)
Andijan massacre	Low	High	Verbal protests	Anger and resentment	No
Annexation of Crimea	Low	High	Verbal protests	Anger and resentment	No (requested)
Attack on Kuwait	Low	High	1991 Gulf War	Anger at Iraqi brutality	Yes (requested)
"Brexit" vote to leave the EU	Very high	High	Resigned acceptance	Mixed but strongly emotional opinions	No
Destruction of al-Kibar nuclear plant	High	Medium	Simmering silence	Disbelief by Iraqi citizens	No
Fukushima nuclear disaster	Low	High	Criticism and distrust of government	Anger and resentment	Low
Invasion of Georgia	Low	High	Request for military aid from the West	Anger and resentment	No (requested)
Quest to find WMD	Medium	Medium	Persistent puzzlement	Skepticism and confusion	No
Trump's foreign security policy	Low	High	Confusion and distrust	Mixed by strongly emotional opinions	No

Table 4.4 Case strategic outcome

Case	Security impact	Short-run effectiveness	Long-run success	Legitimacy image	Unintended effects
9/11 terrorist attacks	High: psychological insecurity and people/property loss	High	Low: rise of ISIS; decline of al-Qaeda	Low	Rise in U.S. patriotism and Islamophobia
Andijan massacre	High: U.S. military bases lost	High	Low: citizen suffering and contagion risk	Low	Strengthening of Uzbekistan-Russia ties
Annexation of Crimea	High: loss of key territory	High	Low: murky Russian periphery relations	Low	Weakened Euro-Atlantic alliance
Attack on Kuwait	Minimal: attack rebuffed	High	Low: Gulf War reverses outcome	Low	Strengthening of U.S.-Kuwait oil ties
"Brexit" vote to leave the EU	High: forced EU disengagement	High	Mixed: continental Europe instability	Medium	Expansion of UK's extraregional ties
Destruction of al-Kibar nuclear plant	High: nuclear thrust thwarted	High	High: risk of continued friction	Low	Increased threat from Iran
Fukushima nuclear disaster	High: mass people/property loss	Low	Low: worsened inside/outside image	Low	Growing opposition to nuclear power
Invasion of Georgia	High: loss of key territory	High	Low: risk of Russian aggression	Low	Russian political isolation
Quest to find WMD	Minimal: U.S. invaded anyway	Mixed	Low: Hussein ousted from power	Low	U.S. credibility loss and intelligence distrust
Trump's foreign security policy	Mixed: friends and foes wary	Mixed	Mixed: risk of future destabilizing jolts	Mixed	Confused and divided policy interpretation

perceived legitimacy of strategic manipulation use is relatively low and usually generated significant consequences affecting the initiator's international relations.

Initiator Manipulation Facilitation Under Information Overload

The cases suggest that, under information overload, transparency and clear understanding may sometimes be undesirable, and enhanced misperceptions based on ambiguity, deception, and surprise may, in limited circumstances such as when direct communication is dysfunctional, serve sound strategic purposes. Sometimes political leaders try to foster cynicism among their citizens, opening them up to conspiratorial interpretations and to uncertainty about the truth, so as to assist in manipulating them toward predetermined political preferences. Significant risks are associated with initiating manipulation, including triggering action-reaction cycles, regional violence contagion, and wider disregard for restraining norms. The global rise of populist nationalist nativist movements increases the potential for successful manipulation of targets, especially in promoting emotional xenophobia or anti-establishment sentiments.

More specifically, manipulation in the cases was often quite effective in the short run (especially militarily) but far less so in the long run (especially politically). Nonstate initiators, exemplified by the al-Qaeda terrorist group in the 9/11 attacks on the United States and by print and digital social media in the Brexit case, were able to use the tools of foreign manipulation just as effectively as state initiators. In cases in which initiators were implicitly or explicitly attempting to overturn the status quo—the 9/11 terrorist attacks, Brexit, the Iraqi attack on Kuwait, and the Russian coercive action in Crimea and Georgia—success rates were just as high as when initiators were trying to preserve the status quo. In cases in which the trigger of the confrontation between manipulation initiator and target was widely recognized, the perpetrators had little idea that it would lead to the problems that it did. Interestingly, although manipulation is often seen as a substitute for force, in most cases initiators combined manipulation with violence.

Target Manipulation Vulnerability Under Information Overload

The cases show that complacency about one's status quo position—often involving wishful thinking—can be deadly in today's rapidly changing world. Even powerful states with huge defense spending and vast intelligence capabilities can be highly susceptible to foreign manipulation. Target ambivalence about resolve often invites such strategic manipulation. Diametrically opposed information-overloaded positions can increase target confusion, and the greater the target confusion, the less coherent its responses and the wider its possible interpretations promoted through manipulation. Crisis time pressure can enhance the impact of strategic manipulation because of the inability of targets to fully cross-check incoming information and to the need to make decisions and take action before thoroughly analyzing their costs and benefits. Many targets, after being manipulated, remained relatively ignorant about strategic ambiguity, deception, and surprise and their implications.

More specifically, although states appear to be the primary manipulation initiators at home and abroad, national governments seem to be just as vulnerable as targets of private manipulation as citizens are as targets of state manipulation. Manipulation that affected not only foreign countries but also internal domestic targets, such as in the cases of the Andijan massacre and the Fukushima nuclear disaster, showed that domestic citizens possess just as much vulnerability to strategic ambiguity, deception, and surprise—with target reactions of anger and resentment, and confusion even more intense—as do foreign governments. In most cases, strategic manipulation targets were at a decided power disadvantage, but in cases in which targets' capacity to respond was high—the United States to the 9/11 attacks, the UK government to the Brexit vote, and Syria to the al-Kibar attack—significant constraints prevented use of available power.

Patterns Specific to Strategic Ambiguity, Deception, and Surprise

The cases suggest that strategic ambiguity is strongly associated with seeing an initiator as unpredictable and discombobulated, even though it can be a manifestation of careful planning. Ambiguity seems to work well when it is used on a highly sensitive target with strong preconceived images that fill in the uncertainty or contradiction gaps provided by the ambiguous communication. In contrast, ambiguity works poorly against determined ruthless opponents able to call one's bluff and respond with brute force.

The cases illustrate that strategic deception can reduce blowback, increase plausible deniability, and complicate attribution. Deception and surprise seem to work better within one's sphere of influence. Revelations of government deception tend to be more stinging if combined with images of government corruption. Strategic deception appears to work poorly when focusing on highly suspicious targets not trusting of any kind of incoming information. Deception also has problems when intelligence on the designated target is sufficiently poor that credible sources trusted by a target cannot be found or manipulated to the initiator's advantage.

The cases reveal that unforeseen events can stymie even experts in many situations—"any ideas that surprise could no longer be achieved in a world of spy satellites and high-technology surveillance equipment must surely have been dispelled."[1] Because surprise often derives from expecting the status quo to continue, it seems to work well if a target is truly wedded to the status quo and is totally unprepared for any sort of dramatic change. Strategic surprise can be as much the product of target ineptitude—especially when overconfident, uninformed, confused, and insensitive—as of skillful initiator manipulation. Surprise seems to fail to provide initiator advantages when targets react quickly and resiliently to unforeseen developments.

Evident Alliance Tensions

One notable security pattern evident in most of the case studies is that tensions emerged between offensive manipulation initiators and their international political allies. This international relations impact resulting from the use of strategic ambiguity, deception, and surprise is rather unexpected, given the prevalence and tacit acceptance of such manipulative activity in a multicultural, anarchic world characterized by information overload.

The international alliance tensions created by strategic manipulation in the ten cases were remarkably parallel. In 2017, American president Donald Trump's use of strategic ambiguity—incorporating both vague and contradictory statements—in foreign security policy caused United States allies to wonder if the United States would keep its established commitments in the future. In 2016, the Brexit vote to leave caused the EU to wonder what it could expect from the United Kingdom in the future, significantly weakening the regional organization, and it pushed Britain away from tight European ties to the wider global Anglosphere. Because of outside perceptions of naked international aggression by a vastly superior military power, the 2014 Russian annexation of Crimea and the 2008 Russian invasion of Georgia further isolated Russia from the international community, worsened Russian relations with its surrounding former satellite states who felt heightened insecurity, and even increased tensions among Western allies who were not of one mind in their defensive response to Russian aggression. In 2011, because of deception involving omission of key details, the Fukushima nuclear disaster increased global distrust of information emanating from the government, along with escalating Japanese citizens' distrust in their own regime. In 2005, because of blatant and brutal human rights violations, the Andijan massacre eroded relations between Uzbekistan and the United States, pushing Uzbekistan more under Russia's sphere of influence and panicking Uzbekistan's neighbors, who feared regional contagion of violence. In 2002–2003, the nondiscovery of weapons of mass destruction in Iraq prior to the Iraq War hurt the relationship between the United States and its allies because they felt it had provided a false rationale for the war and because they felt they could no longer completely trust the accuracy of findings from American intelligence agencies. In 2001, tensions increased between the United States and its Middle East allies from which the al-Qaeda terrorists originated, at least in part because Americans blamed some previously friendly Middle Eastern regimes, most notably Saudi Arabia, for not having taken steps to prevent the emergence of new terrorist cells and to apprehend local terrorists before they could wreak global havoc. In 1990, the Iraq invasion of Kuwait alienated Iraq from its neighbors (with the exception of Yemen), and the American military response pushing Iraq out of Kuwait increased the resentment of many Arab "allies" toward the United States. Unsurprisingly, the strained alliance system failed to ensure external help to manipulation targets in times of need: with the exception of the 9/11 attacks and the Iraq attack on Kuwait, no significant outside aid was forthcoming even when specifically requested.

From a global security standpoint, complexities surround normatively evaluating manipulation-induced alliance tensions due to strategic manipulation. Changing alliances are part of the normal ebb and flow of international relations; sometimes discarding old allies and finding new ones can be quite fruitful in pursuing national interests, and sometimes sharply highlighting differences within an alliance can productively promote informed compromise decisions that make the alliance stronger. Notably, strategic manipulation cannot effectively be cordoned off from the web of global relationships, and international resentment and friction can still result even from engaging in behaviors that almost everyone undertakes on a regular basis.

Global Data Shock Conditional Analysis

Although the case studies suggest that information overload can promote strategic manipulation, and that strategic ambiguity, deception, and surprise can erode global security, these relationships are not universal. The need for careful situational analysis about success and failure associated with strategic ambiguity, deception, and surprise is underscored by the fundamental question "Why do certain manipulations succeed and others do not when the same quantity and quality of relevant information are at the actor's disposal?"[2] Based on the case study patterns, this analysis reveals (1) when information overload most promotes strategic manipulation, (2) when initiators' offensive manipulation and targets' defensive response are most effective, (3) when strategic manipulation is most legitimate, and (4) when strategic manipulation is most dangerous. The circumstances identified point to the need for policies attempting to coherently manage manipulation not to be sweeping but rather to be flexible enough to operate differently in different situations.

Given the limited number of case studies, the conditions identified are inescapably highly tentative. Rather than being viewed as definitive findings, the circumstances identified are better characterized as probable hypotheses derived from the case evidence, needing to be further tested and refined by future research. What is presented here provides important clues for both offensive manipulation initiators and defensive response targets about when their manipulation and countermanipulation strategies might work best. Given that some of the circumstances identified are not fixed but rather susceptible to alteration, manipulation management faces an important choice about whether to try to address propitious background circumstances or the strategic manipulation itself.

When Information Overload Most Promotes Strategic Manipulation

Although information overload generally facilitates strategic ambiguity, deception, and surprise, the cases suggest circumstances when information overload seems particularly likely to promote strategic manipulation:

- When few reliable fact-checking sources exist to evaluate and validate intentionally or unintentionally fake or distorted news stories
- When significant disagreement exists even among credible sources, often due to conflicting information releases, about what is transpiring
- When multiple competing yardsticks are used by different sides to measure relevant security outcomes of manipulation
- When it is extremely difficult to prove whether claims are true because of their inherent vagueness and imprecision or to verification roadblocks

- When the flood of information in a fast-moving crisis is full of noise, accelerating, and constantly changing in a way that defies intelligent filtering

- When media—including traditional print and broadcast media and Internet-based social media—have no enforced rules about what can be reported

- When emotional rhetoric and outlandish claims dominate the public and private discussion about the ongoing security predicament

- When the manipulation target has become so cynical that it no longer has faith that objective truth about the ongoing security predicament really exists

- When both the manipulation initiator and the manipulation target are victims of significant security misperceptions regarding the ongoing security predicament

- When authorities keep changing their assessments and recommendations over time in seemingly haphazard ways about the ongoing security predicament

Several of these identified circumstances are becoming increasingly common, much to the chagrin of intelligence analysts and members of the mass public who are trying to make heads-or-tails judgments about ongoing developments. Moreover, because of frequent negative reactions to too much information, even if an authority is earnestly attempting to do the right thing by regularly updating public estimates, warnings, and guidelines, the effort can backfire and reduce its credibility. Together these information conditions conducive to manipulation promote considerable befuddlement, impeding accurate interpretation of the ongoing predicament and creating vulnerability to manipulation among potential targets.

Specific cases highlight identified conditions. The availability of few reliable fact-checking sources to deal with fake or distorted news was evident in the 2017 Trump foreign security policy case, as sources claiming to do fact-checking often disagreed on what claims were true and false. Source disagreement due to conflicting government information releases was evident in the 2014 Russian annexation of Crimea and the 2008 Russian invasion of Georgia; the media transmitted conflicting information in the 2011 Fukushima nuclear disaster; and conflicting information emerged from perpetrators and victims in the 2005 Andijan massacre. Regarding the use of multiple competing outcome yardsticks by different sides, after the 1990 Iraq attack on Kuwait, Saddam Hussein viewed the outcome as a victory for Iraq because he was still in power, while the United States viewed the outcome as a defeat for Iraq because it was pushed out of Kuwait. As to the difficulty of proving claims because of vagueness or verification roadblocks, the 2017 Trump foreign security policy case illustrates the vagueness obstacle, and the 2002–2003 nondiscovery of weapons of mass destruction in Iraq case illustrates the verification obstacle. The interfering presence of noise was evident in the 2008 Russian invasion of Georgia, the 2007 Israeli destruction of the Syrian nuclear plant, the 2005 Andijan massacre, the 2001 9/11 terrorist attacks, and the 1990 Iraqi attack on Kuwait. Nonbroadcast media lacked enforced reporting rules in the 2016 Brexit case. Emotional rhetoric was present in the 2016 Brexit case and the 2014 Russian annexation of Crimea. With respect to high target cynicism doubting the existing of truth, Putin pushed this exact objective in the 2014 Russian annexation of Crimea. Mutual initiator-target misperceptions were evident in the 2008 Russian invasion of Georgia and the 2002–2003 non discovery of WMD in Iraq. Finally, confusing changing government assessments over time were evident in the 2011 Fukushima nuclear disaster.

When Strategic Manipulation Is Most Effective

The cases also suggest circumstances under which policy effectiveness appears to be most likely to result from offensive use of strategic manipulation and the defensive responses to it. Initiators' offensive use of strategic ambiguity, deception, and surprise seems likely to be most effective under the following conditions:

- When an initiator attempts to blame its problems on foreigners and to distract citizen and international community notice away from its failings

- When an initiator attempts to portray to citizens and the international community aggressive actions as defensive within its sphere of influence

- When an initiator attempts to confirm among targets strongly held preconceived images about deeply held security fears and identity aspirations

- When an initiator attempts to deter and disorient an adversary or keep it from taking any provocative steps, particularly during a heated conflict

- When an initiator attempts to affect targets whose leaders are characterized by complacency, wishful thinking, and predictable static responses

- When an initiator attempts to affect targets whose leaders are ambivalent about their resolve or confused by opposing arguments about an issue

- When an initiator attempts to affect targets whose leaders perceive that they are operating under extreme time pressure regarding an issue

- When an initiator attempts to affect targets who understand relatively little about the dynamics of manipulation and how to respond to it

In contrast, targets' defensive response to strategic ambiguity, deception, and surprise seem most likely to be effective under the following conditions:

- When a target attempts to respond by seeking and receiving advice and assistance from capable, prudent, and experienced outside parties

- When a target attempts to respond to a brash initiating leader displaying consistent overconfidence, inattentiveness, and intransigence

- When a target attempts to respond to a pariah initiator widely scorned by the international community

- When a target attempts to respond while possessing coherent goals and strategies, with unified support within the government and citizenry

- When a target attempts to respond while most people in its society are vigilant, prepared, adaptive, and resilient when facing unexpected change

- When a target attempts to respond while possessing tangible attributes or resources deemed centrally important to major powers

- When a target attempts to respond while possessing overall military, economic, and political power superior to that of the initiator

- When a target attempts to respond while it has credibly isolated through intelligence an exploitable initiator vulnerability

Propitious circumstances for initiators' offensive manipulation can play off either initiator strengths or target weaknesses. All circumstances conducive to policy effectiveness are unlikely to be present at the same time for either initiators or targets. A case review reveals that far more manipulation initiators than manipulation targets encountered conditions for effective management.

We look first at cases highlighting initiators' offensive use of manipulation. Blaming problems on foreigners is exemplified by the 2014 Russian annexation of Crimea, and distracting others from initiator failures is exemplified by the 2005 Andijan massacre and the 2011 Fukushima nuclear disaster. Portraying aggressive action within one's sphere of influence as defensive is illustrated in the 2014 Russian annexation of Crimea, the 2008 Russian invasion of Georgia, and the 2005 Andijan massacre. Using manipulation to confirm targets' deeply held security fears and identity aspirations (especially populist nativist sentiments) is found in the 2017 Trump foreign security policy case (fear of national decline) and the 2016 Brexit case (fear of illegal migrants). Deterring an adversary by preventing it from taking provocative steps is evident in the 2007 Israeli preemptive destruction of the Syrian nuclear power plant. Choosing targets with complacent leaders—ones wedded to the status quo—is shown in the 2001 9/11 al-Qaeda terrorist attacks on the United States. Undertaking manipulation when target leaders have ambivalent resolve (as opposed to ruthless feisty antagonism) is present in the 2016 Brexit case. Dealing with targets whose leaders feel under extreme time pressure is displayed in the 2002–2003 nondiscovery of weapons of mass destruction in Iraq, because UN inspectors had a narrow window in which to find WMD in the country. Last, selecting a target understanding little about manipulation and how to respond to it was characteristic of Kuwait in the 1990 Iraqi attack on the country.

We look next at cases highlighting the effectiveness of targets' defensive response to manipulation. Georgia made the proper move by requesting major outside assistance after the 2008 Russian invasion, despite the understated response from the West. As to taking on an initiator with a brash overconfident leader, that is exactly what occurred with respect to Saddam Hussein in the 2002–2003 WMD in Iraq case and the 1990 Iraq attack on Kuwait. Regarding possessing coherent goals and strategies with unified support, as well as having overall power superior to that of the initiator, the American response to the 9/11 terrorist attacks is illustrative. Last, having a strategically valuable resource is exemplified by Kuwait's possession of oil in the 1990 Iraqi attack on the country.

As expected, it seems that strategic ambiguity, deception, and surprise will be most effective when used by savvy initiators—with experience in successful manipulation and accurate information on targets' communication filters, cultural predispositions, and value interpretation—against unsavvy targets that are not expecting them and have no effective warning or intelligence system to help them know how best to respond once they become victims of foreign manipulation. Regardless of their overall level of political, economic, or military power, the absence of a highly vigilant defense system familiar with how to spot and how to respond to strategic ambiguity, deception, and surprise would create for potential targets the global security equivalent of a sitting duck for foreign manipulation. Strength in using manipulation and defending against it is decidedly separate and different from traditional political, military, or economic strength, as often these manipulation techniques can be effectively used in asymmetric confrontations to compensate for deficiencies in traditional elements of power.

When Strategic Manipulation Is Most Legitimate

Even though usually outsiders see foreign manipulation as illegitimate—a pattern confirmed in most cases studied—in some circumstances strategic ambiguity, deception, and surprise could most likely be perceived as legitimate (in the views of domestic citizens and the international community):

- When the use of ambiguity, deception, or surprise occurs within the initiator's sphere of influence
- When blind trust by the target in the manipulation initiator has traditionally been extremely high
- When the use of ambiguity, deception, or surprise is coherent and understandable over time
- When onlookers are very cynical and consequently have a low bar for what might be considered just
- When the use of ambiguity, deception, or surprise does not blatantly violate national or international law
- When the use of ambiguity, deception, or surprise does not directly cause any loss of life or property damage
- When domestic and international media fail to play an effective watchdog role in reporting on the manipulation
- When any outside sanctions taken by upset outsiders against a manipulation initiator backfire
- When no widespread feeling of helplessness is present in a target after the use of ambiguity, deception, or surprise
- When the target and its allies do not choose to vociferously protest the use of ambiguity, deception, or surprise

Sometimes global failure to react strongly to the illegitimacy of foreign intervention is a function of lack of international awareness, as exemplified by the 2005 Andijan massacre taking place in a remote location or the 2007 Israeli attack on Syria eliciting silence from the victim. Perceived legitimacy may not be absolutely vital for manipulation success within a global anarchic setting, but acceptance of behavior can nonetheless play an important role—particularly if it results in the lifting of onerous sanctions—in determining whether a foreign manipulation initiator feels empowered to take similar actions again with impunity in the future.

When Strategic Manipulation Is Most Dangerous

While employing ambiguity, deception, and surprise is always risky, the cases suggest circumstances under which strategic manipulation seems likely to be particularly dangerous to global security (in terms of scope and intensity of damage, lack of preparation to deal with consequences, and possible backfire effects). The conditions when strategic foreign manipulation seems most dangerous are as follows:

- When a target's awareness of an initiator's manipulation is combined with the target's perception of the initiator's corruption
- When the quality of intelligence is so poor that the risks of actions are not well understood, resulting in hazardous behavior
- When an ongoing security predicament is deteriorating and initiator manipulation is interpreted as denial of this reality and refusal to take remedial action
- When a manipulation initiator has a long tradition of open transparent communication with citizens and allies who firmly expect and desire this pattern to continue
- When initiator manipulation dramatically erodes an international organization of which it is a member and whose membership it deeply values
- When, during a major crisis, a manipulation initiator withholds or distorts vital information, causing tangible loss of life and property
- When an externally dependent manipulation initiator sufficiently alienates the international community that it finds itself politically isolated
- When because of unsecured leaks, the true motive and nature of an intended manipulation goes public before the manipulation can be implemented

- When information manipulation leads to a decision that would never have been undertaken otherwise that severely harms affected societies
- When the confusion engendered by information manipulation is so deep that policy paralysis occurs when urgent action is needed
- When the obfuscation created by information manipulation prevents identification and punishment of the perpetrators of significant security-degrading actions
- When information manipulation produces such profound intolerance of another group or state that it leads to conflict and violence

Notably, while manipulation gains are often transitory, the aftereffects of discovered manipulation tend to persist. Few societies can remain stable indefinitely if the mass public is seething with anger, resentment, and confusion about government behavior at home and abroad.

Specific cases highlight these dangers. Exemplifying international awareness of both manipulation and corruption are the 2014 Russian annexation of Crimea (the Putin regime was seen as corrupt); the 2011 Fukushima case (the Japanese government and the TEPCO nuclear power company were seen as corrupt); and the 2005 Andijan massacre case (the Uzbek government was seen as corrupt). Illustrating low-quality intelligence leading to hazardous behavior, in the 1990 Iraq invasion of Kuwait Saddam Hussein's intelligence on the American reaction was so poor that he was shocked at the 1991 Gulf War. Deteriorating conditions caused manipulation to be viewed as denial of this reality—in terms of citizens' and foreigners' view of the Japanese government—in the 2011 Fukushima nuclear disaster. The image of open transparent communication being tarnished occurred in both the 2017 Trump foreign security policy case and the 2016 Brexit case. International organization erosion was painfully evident in the 2016 Brexit decision to leave the EU. Withholding or distorting vital information leading to loss of life and property occurred both in the 2011 Fukushima nuclear disaster and in the 2005 Andijan massacre. Manipulation alienating the world enough to create political isolation occurred with Russia in the 2014 Crimea annexation and the 2008 Georgia invasion. Unsecured leaks undoing a manipulation before it takes effect have occurred within Donald Trump's administration, limiting his ambiguity and surprise policy's effectiveness. As to manipulation leading to policies that would not have been undertaken otherwise, including creating intolerance and violent conflict, for the United States the al-Qaeda manipulation behind the 9/11 attacks led to the

2001 war on Afghanistan, and the WMD kerfuffle with Iraq led to the 2003 Iraq War. Confusion leading to policy paralysis is shown by NATO hesitation in response to the 2014 Russian takeover of Crimea. Last, manipulation preventing perpetrator punishment is evident in the 2005 Andijan massacre.

Notable Case Lessons

For strategic initiators seeking better offensive manipulation outcomes, the road to success may not be as smooth and black-and-white as it first appears. Even when strategic use of ambiguity, deception, and surprise works well, this manipulation may not be sufficient by itself to attain all desired security goals. As the 2008 Russian invasion of Georgia case shows, even though Georgia won the propaganda battle to garner global support and sympathy, that victory was insufficient to translate into significant tangible Western support and overall foreign security policy success. As evidenced in many of the cases studied, short-term military success may fail to attain long-term political objectives. The question of sustaining success over time is accentuated given omnipresent possibilities of leaks concerning one's true precise intent toward targets. So even though manipulation is generally inexpensive and operates well in today's information-overloaded anarchic global setting, manipulation initiators need to think carefully about when, where, and how to apply strategic ambiguity, deception, and surprise in the current security environment.

Moreover, strategic initiators may find themselves in a consistency bind when applying ambiguity, deception, and surprise. If these initiators attempt to garner internal and external credibility by being extremely consistent over time in their statements and actions, they can be viewed by key constituencies as out of touch, unresponsive, and refusing to adapt to changing circumstances. If, on the other hand, these initiators attempt to garner internal and external credibility by constantly updating and changing over time their statements and actions to cope with changing circumstances, as the Japanese government did during the Fukushima crisis, then they can be viewed by key constituencies as inconsistent, irrational, capricious, haphazard, and incompetent. Either way the risk is losing not only credibility but also respect, trust, and future manipulation potential. Although the domestic public, international allies, and outside observers may generally believe that consistency—like transparency—is always a foreign security policy virtue, in reality it has key deficiencies in a rapidly changing world, and in many ways within a world of cultural diversity, political anarchy, and information overload, it is just as vulnerable to misinterpretation as is strategic ambiguity, deception, and surprise.

For manipulation targets seeking better defensive responses, especially for potential major power targets, traditional security-enhancing measures generally fail when facing sophisticated foreign manipulation: for example, tangible coercive ways of preventing or punishing outside interference—including military and economic sanctions such as those imposed against the Russian government in the Crimea case, the Uzbekistan government in the Andijan massacre case, and the Iraqi government in the cases of WMD and Kuwait invasion—do not work well to forestall intangible, nuanced, and subtle foreign-initiated strategic ambiguity, deception, or surprise. Remaining relatively complacent about deterrence or defensive response systems aimed at foreign manipulation, evident especially in by the American government in the 9/11 terrorist attacks, is never wise given rapidly evolving techniques of strategic ambiguity, deception, and surprise and the appeal of using manipulation in today's information-overloaded world. As a result, countries that are most vulnerable to outside manipulation need to have robust contingency plans, notably absent for manipulation targets in virtually all of the cases, for rebounding from undesired jolts, confusion, and uncertainty. Since "timely, unambiguous warning is nice to have" but cannot be counted on,[3] manipulation targets cannot base such multifaceted contingency response plans on having any significant preparation time for their implementation.

The Countermanipulation Conundrum

This study's findings about how strategic manipulation functions under information overload could lead to depressing conclusions about the potential for coping with global data shock. Offensively, information overload can enhance strategic manipulation's success and versatility to operate in different situations. Despite their low perceived legitimacy by the international community, their escalation of international alliance frictions, and their lack of long-run staying power in achieving political objectives, the uses of strategic ambiguity, deception, and surprise appear to be proliferating today in international relations without encountering significant internal or external roadblocks.

Defensively, the information explosion can hog-tie target state officials into feeling that they lack a sound basis for making complex foreign security policy choices about minimizing manipulation because of the absence of perceived definitively confirming valid and reliable intelligence data. Even for those working full-time on data interpretation, it seems harder than ever to isolate crucial signals out of all the noise because contradictions seem too stark, persistent knowledge gaps too large, and data credibility too low or uncertain.

However, as the next chapter demonstrates, with the debilitating threat of incoherent security policy and global chaos at stake, there is hope for forward progress in preventing or responding to manipulation. Instead of wishful-thinking-enhanced complacency about the status quo and the sufficiency of existing countermanipulation defenses, more diversified preparation, vigilance, and contingency planning can improve foreign manipulation early-warning systems. Instead of traditional flawed security measures overwhelmed by sophisticated foreign intrusion efforts, more innovative, informed, and sensitive target responses can succeed in constraining foreign manipulation's disruptive impact.

5 Managing Global Data Shock

THE CASE STUDY PATTERNS suggest ways to help manage information overload and to help both initiators and targets to manage strategic ambiguity, deception, and surprise. Success in these areas is not simple for either offensive initiators or defensive targets, despite the earlier note that offense so far has the advantage over defense. Within a world filled with global cultural diversity and international political anarchy, there is a pressing need to find ways to develop incredibly deep sensitivity to others despite facing very different societies with very different values, beliefs, and modes of expression. Within a world filled with rapidly changing circumstances, there is a pressing need to find ways to achieve adaptability and nimbleness despite the limitations posed by human cognitive frailty and organizational decision inflexibility. Within a world filled with high data quantity and low data quality, there is a pressing need to find ways to properly sift through information to discover what is really important and relevant despite the dangers of complacency, confusion, and exasperation. Because of the high stakes and low error tolerance, little opportunity exists for trial-and-error experimentation with what might work to achieve desired foreign strategic manipulation outcomes.

Creative thinking is vital to cope with foreign data interpretation and strategic manipulation, including combining fluid, innovative, and responsive measures, avoiding stick-in-the-mud repetitive use; discovering or creating new information and communication channels; and engaging in more systematic advanced contingency planning. It is preferable to operate within adversaries' preexisting beliefs, with improved cross-cultural sensitivity and wariness about projecting one's own assumptions onto others. Rigorous analytical methods must root out false or misleading data, reduce overly simplistic data interpretation, improve recognition of one's own biases, fill in knowledge gaps, and place data assessment in a more meaningful context. It is critical to encourage vigilance and resiliency, striving to improve (1) detection systems while adversaries accelerate their stealth and (2) signal isolation while noise proliferates. Identified threats and vulnerabilities require projection well into the future so as to forestall counterproductive developments. Past effort evaluation requires gathering more comprehensive inside and outside feedback about outcomes, minimizing overconfidence while maximizing receptivity to criticism and eagerness to adapt, and improving learning to refine future behavior. Ultimately, it appears most crucial to avoid the temptation of fatalistically accepting the unintelligibility of information overload and the unmanageability of strategic manipulation.

Given the high global frequency of confusion, failure, and backfire effects in managing information overload and strategic manipulation, there is a lot of room for improvement. So before establishing management strategy priorities and introducing remedial management recommendations for global data shock, this study first reviews deficiencies in the common ways manipulation initiators and targets have attempted so far to manage information overload and strategic ambiguity, deception, and surprise. Flawed standard operating procedures have persisted partially because of (1) insufficient review of their effectiveness and (2) concerted advocacy for them by proponents. Despite intelligence analysts' increasingly earnest efforts and new computerized analytical approaches' potential to dramatically ameliorate data interpretation and resulting policy decisions, the payoff has not yet been fully evident. So it is important to identify and root out dysfunctional attitudes and behavior among political leaders, organizational structures, and planning processes.

Global Data Shock Mismanagement

The record in managing information overload and strategic manipulation has reflected key deficiencies, summarized in Figure 5.1. Information overload management deficiencies include assuming that information overload interpretive challenges can be overcome through introducing additional data, applying proper filtering techniques, going offline, and looking simply at what most sources say. Initiator offensive manipulation management deficiencies include using manipulation only as a last resort, focusing exclusively on short-term rather than long-term security consequences, abandoning approaches that do not initially work well, committing to a single manipulation technique,

INFORMATION OVERLOAD MISMANAGEMENT

Assuming that interpretive challenges can be overcome through the introduction of additional "facts" or empirical studies' findings
Relying exclusively for data interpretation on data filtering, lacking independent objective means to determine data credibility
Resorting to going offline, a decidedly "cutting off your nose to spite your face" approach, to managing information flows
Accepting as truth what is agreed on by the largest number of sources, especially if it supports one's preconceived ideas of what is true

INITIATOR MANIPULATION MISMANAGEMENT

Using manipulation inappropriately only as a last resort—in back-against-the-wall, power-deficiency, or out-of-options situations
Focusing on short-run gains and ignoring long-term security consequences
Responding to failure by throwing in the sponge instead of tweaking failed techniques to improve their success
Putting all of one's eggs in one basket instead of maximizing options to overload target analysis and response capacities
Using a scattershot startegy, trying everything indiscriminately to see what works instead of prudently selecting the best options

TARGET RESPONSE MISMANAGEMENT

Reviewing the last instance of foreign manipulation in ways that do not provide immunization from future manipulation
Employing deflection—denying manipulation importance, responsibility for failure, or ability to predict, prepare, or respond
Trusting that persuasion is sufficient to deter adversaries from engaging in manipulation
Being totally committed to a single defensive response that is blindly believed to work in all situations
Relying exclusively on the most overconfident, cognitively consistent internal experts for interpreting signals and suggesting responses

Figure 5.1. Global data shock mismanagement

and randomly trying every option to expose target weaknesses. Defensive manipulation response deficiencies include deriving the wrong kinds of learning from past intrusions, deflecting responsibility for failures, trusting persuasion alone to deter enemy manipulation, committing totally to a single defensive response, and relying too exclusively on seasoned internal experts to interpret incoming signals and recommend responses.

Information Overload Mismanagement

Perhaps the most basic deficiency in coping with information overload is to assume that interpretive challenges can be overcome through introducing ad-

ditional "facts" or empirical findings, resolving ongoing into heated foreign security debates or convincing contending sides to modify their firmly held positions. For example, often following intelligence failures,[1] "the reaction of the
intelligence community to many problems is to collect more information, even
though analysts in many cases already have more information than they can di-
gest."[2] Given global information overload and omnipresent foreign communication manipulation, this appears to be a decidedly faulty premise. A standard
intelligence maxim is that "facts don't speak for themselves";[3] key differences
not only in data interpretation but also in deciding the significance of data and
their relationship to broader security trends can prevent consensus. Moreover,
the distinctive interpretive values among contending parties may cause them to
process information quite differently and assign very different credibility to the
same data and sources. Beyond a certain point, focusing on acquiring additional information can at best encounter diminishing returns and at worst be
totally useless or counterproductive. Indeed, knowledge "does not guarantee
enlightenment or wisdom."[4]

Another inadequate common solution to information overload is based
on the contention that applying proper filtering techniques is sufficient to help
people to see the truth:

> What can be done about information overload? One answer is technological:
> rely on the people who created the fog to invent filters that will clean it up. Xe-
> rox promises to restore "information sanity" by developing better filtering and
> managing devices. Google is trying to improve its online searches by taking into
> account more personal information.[5]

Indeed, most strategies for coping with information overload "in essence boil
down to filter and search"—"when information is cheap, attention becomes ex-
pensive."[6] Despite the theoretical utility of filters, they are grossly inadequate
in practice. Clever data manipulators continually discover new ways to circumvent or overcome filters by disguising the nature of sources, creating un-
orthodox transmission channels, and taking advantage of a widening variety
of fast-changing real and fake data. Even the best artificial intelligence filtering
algorithms cannot keep up with such adept adjustments. Moreover, no matter
how much filters narrow down useful information or appear to separate wheat
from chaff, filters possess no independent objective means of assigning credibility to questionable data (instead, they tend to rely on user preferences) and
cannot improve interpretation for data deemed important (instead, they tend

to rely on user judgments). Finally, "smarter filters cannot stop people from obsessively checking" their computing devices for more and more information, communication, and updates—some people "do so because it makes them feel important; others because they may be addicted to the 'dopamine squirt' they get from receiving messages."[7] Such challenges are magnified for foreign security issues, in which subtle adversary uses of strategic ambiguity, deception, and surprise often elude any form of screening.

A third defective response to information overload entails going offline. The underlying logic is simple: if you disconnect from the Internet, you will cease to be bombarded with so much information that you cannot process it, giving you more time to analyze and reflect on data you already have. Terminating online access can be machine-automated—for example, "a popular computer program called 'Freedom' disconnects you from the web at preset times"—or alternatively triggered by human willpower, exemplified by when you voluntarily "ration your intake" or "turn off your mobile phone and internet from time to time."[8] However, this "cutting off your nose to spite your face" approach purposely shuts down all digital information access, making it easy to miss vital data during rapid global change. If an organization rejects cutting off Internet access, such "self-discipline can be counter-productive," as "some bosses get shirty if their underlings are unreachable even for a few minutes."[9] Because of growing digital data dependence, there has been a decline in willingness to disconnect even temporarily from the Internet information pipeline.

A final faulty response to information overload involves accepting as truth what appears in the largest number of sources, particularly when "so much evidence supports your preconceived idea" of what is true.[10] Such a simple but misguided fix is based on the assumption that if enough covert and overt sources agree with each other, the information must be accurate. However, today such volumetric-based assessments are easily manipulated by flooding all sorts of outlets with distorted data that appear (falsely) to be from completely independent sources.

Initiator Manipulation Mismanagement

It is quite common to hear that strategic ambiguity, deception, and surprise are best used only as a last resort—if a party's back is to the wall, if the party is weak or outmatched, or if the party has already tried everything else with no success. However, this conclusion seems highly inappropriate—"strategists should not view deception [and other manipulation techniques] as a tool of last resort for the weak, but as an instrument to gain or maintain a strategic advantage over a rival."[11] Despite the need for caution in using manipulation, early action can often be better than waiting until one comes under attack, since going second can be disadvantageous in allowing adversaries to prepare for retaliation—a first strike by an opponent could be disarming and severely limit or even prevent any response. Regarding the 2007 Syrian construction of a nuclear reactor, had Israel waited to use forceful strategic surprise until diplomacy failed, effectiveness might have been minimal. Notably, "the great strategic lesson of the 1930s is that early military action is far more preferable than a last-resort use of force against that very rare, powerful enemy who is both politically unappeasable and militarily undeterrable."[12] In time-pressured foreign policy crises, initiators cannot wait long to respond, even if this speed means short-circuiting a full cost-benefit analysis of all available options and their consequences.

Another common flawed mode of manipulation initiation is based on the premise that because the primary benefit is short-run, long-run security impacts can be downplayed. While such benefits do tend to be only short-run, it is a mistake for manipulation initiators to avoid planning to manage long-run security consequences. The 2011 Fukushima nuclear disaster and the 1990 Iraqi attack on Kuwait illustrate how the failure to plan for long-run negative responses (domestic in the Fukushima case and international in the Iraq case) negated any short-run gains.

When a strategic manipulation failure occurs, initiators often simply decide not to reuse the technique applied, at least against that type of target. The simple logic behind quick option abandonment is that if one approach does not work, then another should be tried. However, this ignores considering refined versions or creative variants of past unsuccessful manipulation techniques. Given the subtle complexities in foreign strategic manipulation, initiators need to be more patient, not expecting immediate success on the first try, and work to improve with each successive new thrust, learning from past mistakes what specific elements work best and worst.

Sometimes offensive manipulation initiators succumb to an "all eggs in one basket" approach—committing to a single ambiguity, deception, or surprise strategy with all available resources to maximize chances of success. President Donald Trump's obsessive emphasis in on unpredictability as the sole key to foreign security policy success illustrates the need to consider a wider range of manipulation techniques. Being too wedded to a single preferred strategy makes it too easy in many circumstances for manipulation targets to anticipate and cope with it.

An opposite faulty manipulation approach is the scattershot strategy—trying everything indiscriminately to see what works, instead of prudently selecting the best option after comparative analysis. The premise here is that since one cannot decisively determine in advance what works best, one might as well just randomly try every option available in the hope that at least one sticks and succeeds in attaining desired goals. However, this chaotic approach makes little sense because of limited resources, the need for credible commitment, and the high costs of failed manipulation. For example, al-Qaeda's deception manual recognizes this limitation by recommending that for success "all falsification matters be carried out . . . not haphazardly."[13] When operating in secret, sensitive, and protected areas, proceeding randomly is especially prone to fail.

Target Response Mismanagement

A common flawed target solution is to review the last foreign manipulation success to learn how to prevent similar future errors. However, often the wrong kinds of learning take place, at least in part because foreign manipulation methods usually change dramatically over time:

> Throughout history, superpowers and small nations, democracies and authoritarian regimes, have surprised and deceived each other, and fallen victim to deception and surprise, often with devastating results. Yet, as the collected repository of human wisdom, either history has not been a very good teacher or we have not been very good students. Exposure to hostile deception does not provide immunization from being deceived again, either by the same actor who "got you" before, or by someone else.[14]

The manipulation pattern is dynamic, as most initiators vary the kinds of manipulation they use against a particular target, so preparation based just on the last type of manipulation encountered may prove misguided. Thus potential manipulation targets' hyperattentiveness to the last instance of having been duped does not make sense. Furthermore, a key problem with this type of learning as a target response to manipulation is that it can easily lead to overcorrection or boomerang effects (following the cognitive conceit pattern) in response to a discovered vulnerability. Moreover, such learning can end up reducing certain vulnerabilities or areas of exploitation while at the same time inadvertently creating new ones.

Especially within government bureaucracies, another problematic target defensive manipulation response is deflection—denying that a foreign manipulation that has just transpired was important, that one's own faulty analysis was responsible for foreign manipulation success, or that anyone could have predicted foreign manipulation, prepared for it, or reduced its negative impact. This deflection serves not only to prevent effective containment of a foreign manipulation's negative security impacts but also to reinforce inappropriately the adequacy of existing standard operating procedures and countermeasures already in place. As is readily apparent, such deflection can prove disastrous, reinforcing complacent acceptance of the status quo and preventing needed target adjustments. Blaming other agencies, groups, or individuals for the vulnerability to manipulation—and implicitly demanding that they, not you, figure out and make necessary changes—reflects dysfunctional scapegoating that never solves problems. For example, it is typical to blame media bias for susceptibility to outside manipulation, often leading to fatalistic acceptance of foreign manipulation as an unchangeable constant to which no defensive response could prove effective.

A related defective but common target manipulation response is to trust that extensive persuasion alone is sufficient to deter adversaries from future offensive efforts. For example, in the 2014 Russian annexation of Crimea, outsiders desiring Russian restraint were overconfident that Russia was playing by their rules and were thus overoptimistic that "it would be possible to persuade Russia" to avoid a coercive solution to the tensions in the region.[15] Similarly, in the 2016 Brexit vote, while "numerous European leaders made a convincing argument that only through unity could the continent escape its bloody past and guarantee prosperity,"[16] that was insufficient to dissuade the "Leave" movement's initiators from pursuing their separatist agenda. This emphasis on exclusively verbally convincing a foreign offensive initiator to stop can reduce credibility by artificially excluding potentially useful coercive options.

Similarly, because of resource constraints or pigheaded belief in its superiority, a prevalent deficient manipulation response is to become so totally committed to a particular defensive response or set of defensive responses that targets blindly believe it will work in all situations:

Don't fall in love with any plan, policy, program, or assessment. Don't expect the opponent to cooperate. Have a branch and sequel to address the unexpected along the lines of "what if?" and "what next?" Pay attention to what both adversaries and allies are saying and doing—especially if there is a mismatch between words and deeds. Don't discount indicators just because they point to things *you* would never do. There are no universal standards of rationality or recklessness.[17]

Often the belief among targets that a particular defensive response will work to thwart manipulation is so strong that its aftereffects will be reinterpreted so as to make the response appear in retrospect as if it had been successful.

Finally, for some manipulation targets, the defensive solution is to rely exclusively on internal seasoned experts to interpret incoming signals and recommend responses. The rationale here is that they are in the best position to have the historical knowledge necessary to know best how to react to incoming manipulation. Unfortunately, even though such internal expert interpretation is important, it can maximize the changes of cognitive consistency and confirmation bias, carrying over past images into the present: such internal experts tend to be the last to spot surprising discontinuous change in predicaments they have studied for decades, quite common in an anarchic global security setting, because "experience suggests that experts in their fields are not necessarily the most likely to recognize and accept changes or new types of data."[18] Internal experts also can sometimes be prone to exhibit overconfidence, when tentativeness would seem more prudent, about knowing the intentions of adversaries engaging in manipulation and about recommending certain responses.

Overall Mismanagement Security Impact

The combination of frequent mismanagement of information overload, offensive manipulation initiatives, and defensive manipulation responses has increased global insecurity and created ominous backfire effects. Too much certainly that one is doing the right thing, often due to postmanagement evaluation failings, can lead to misplaced complacency that everything in the information and communication arenas is fully under control. In the end, barking up the wrong tree in response to global data shock can often be worse than doing nothing at all.

Comparative Prioritization of Management Strategies

Combining the empirical case study analysis with the earlier conceptual discussion, the following insights emerge about contrasting global data shock management strategy priorities:

- To improve isolation of where critical interpretation differences exist, it seems more important to reward intelligence analyses for generating carefully qualified assessments than for making broad predictive judgments based on fluid conflicting evidence.

- To improve the quality of foreign manipulation assessments, it seems more important to reward intelligence analyses that focus on the high-risk route of interpreting adversaries' intentions than on the safer route of simply reporting adversaries' capabilities and actions.

- To improve coping with information overload challenges, it seems more important to focus on finding ways to develop better analytical tools for data interpretation than on finding ways to collect additional data.

- To improve learning over time, it seems more important to focus on detailed investigations of the long-term patterns of circumstances surrounding abysmal failures than of those surrounding sparkling successes.

- To improve surmounting information interpretation barriers, machine-based big data analysis and artificial intelligence algorithms seem much more adept at overcoming human cognitive frailty than security information unreliability.

- To improve signal detection within a noisy information environment, it seems more important to spot rhetoric-reality discrepancies, especially between political statements and military actions, than to develop fixed abstract conceptual rules to determine data utility.

- To improve recognition of disinformation and fake news, it seems more important to focus on a thorough investigation of the validity, reliability, and objectivity of data sources than on whether the information confirms or contradicts previous assessments.

- To improve assessment of manipulation outcomes, it seems more important to focus on whether enemies' attitudes and behavior transform than on whether one receives criticism or praise from friends.

- To improve manipulation initiators' consideration of long-run risks, it seems more important to focus on the potential for local and regional instability than on negative international community reactions.

- To improve manipulation initiators' offensive thrusts, it seems more important to focus on increasing understanding of designated target vulnerabilities than on increasing acquisition of sophisticated manipulation tools.

- To improve manipulation targets' awareness of strategic ambiguity, deception, and surprise used against them, it seems more important to expect a pattern of continual manipulation efforts rather than to classify these as rare isolated incidents.

- To improve manipulation targets' defensive responses, it seems more important to focus on promoting their vigilance and preparedness, heightening post manipulation recovery and resiliency, than on reducing the frequency of strategic manipulation.

Each reflects an important choice by offensive initiators or defensive targets about what seems most critical to emphasize and deemphasize, given limited resources and time, in managing strategic ambiguity, deception, and surprise under information overload.

To improve isolation of where critical interpretation differences exist, it seems more important to reward intelligence analysts for producing carefully qualified assessments than for making broad predictive judgments based on fluid conflicting evidence. In this thrust, it would be especially important for these analysts to pinpoint differing levels of certainty for different facets of overall reports, explaining the basis for such differences, rather than just blanket confidence ratings. More basic training could help intelligence analysts recognize their processing limitations—including human cognitive frailty and organizational decision inflexibility impeding sound data interpretation—in order to reinforce their willingness to be more tentative about conclusions with questionable support. The 2002–2003 nondiscovery of weapons of mass destruction in Iraq by the United Nations and the United States illustrates the importance of this priority.

To improve the quality of foreign manipulation assessments, it seems more important to reward intelligence analyses focusing on the high-risk goal of interpreting adversaries' intentions than on the safer goal of simply reporting adversaries' capabilities and actions. Although collecting objective data on capabilities is easier than those on intentions, intentions intelligence is often more crucial and influential in foreign manipulation assessment. For example, in the 2014 Russian annexation of Crimea, Western policy failure resulted from misperception of Russian intentions; in the 1990 Iraqi attack on Kuwait, the American failure to anticipate the military action resulted from misreading

Saddam Hussein's intentions. Offensive initiators realize that strategic manipulation works better when stressing intangible intentions than observable capabilities primarily because in comparing intentions and capabilities, political and military intentions "are much simpler to conceal."[19] Too often intelligence analysts are tempted to take the path of least resistance and avoid a proper focus on adversary intentions:

> By tacit or unspoken agreement, perhaps because the issue is so "hot," it [the group responsible for warning] does not really come to grips with the crucial question of intent. The end product is a non-controversial, wishy-washy statement such as: "A high level of military activity continues along the XY border, and the situation remains very tense." Naturally, everyone can agree on this—it cannot be said to be "wrong." But not only is this unhelpful to the policy official, it also means that intelligence has not really gone through the rigorous intellectual process of analyzing the meaning of the available data and attempting to interpret its true meaning.[20]

Often this reluctance to focus on intentions is quite understandable—"the idea that intelligence either cannot or should not be making judgments of intentions usually arises only when there are sudden or major changes in a situation requiring a new judgment, and particularly a positive judgment about whether aggressive action may be planned."[21] While the descriptive reporting of visible capability trends may minimize the chances of being called out for making a huge analytical mistake, it does not assist much in the absolutely crucial warning about foreign manipulation.

To improve coping with information overload challenges, it seems more important to focus on finding ways to develop better analytical tools for data interpretation than on finding ways to collect additional data. Often discovering new analytical tools useful for managing manipulation involves thinking outside the box well beyond conventional wisdom and standard operating procedures to cope with ambiguity, deception, and surprise. Improving interpretation of existing information usually entails digging much deeper than usual into the historical, political, and cultural context, a task that becomes truly unwieldy if the focus is on gathering more and more data. Collecting additional data does little when one already has too much information, and in any case personal biases and organizational inertia would usually tend to just fit the new information into preexisting images. For example, given strong emotionally charged positive and negative reactions to his presidency, collecting more information on American president Donald Trump's intentionally ambiguous and contradictory foreign security policy statements might serve little purpose in being able to accurately predict future policies.

To improve learning over time, it seems more important to focus on detailed investigations of the long-term patterns of circumstances surrounding abysmal failures than of those surrounding sparkling successes. Emphasizing learning from successes rather than failures increases the chances of inappropriate carryover into the present of past assumptions that no longer apply because of changing conditions. Often failure to anticipate, failure to learn, and failure to adapt are central impediments to effective manipulation management.[22] Initiators should examine carefully the manipulation aftereffects to see exactly where and why the manipulation went wrong in order to tweak its application so as to be more successful in future efforts. The resulting "awareness of the obstacles to unbiased information processing may not only stimulate decisionmakers to avoid biases but also encourage them to institute basic training to improve information processing."[23] In particular, "much more effort than is now expended should be devoted to preparing decisionmaking systems to expect and manage the potential negative consequences of misperceptions"[24] associated with such intelligence failures. Organizational cultures need to transform such that those who admit mistakes and a willingness to reexamine the causes elicit praise rather than humiliation from their peers and superiors. Moreover, the maintenance and nourishment of fluidity in strategies and responses—and openness to the unexpected—is essential in managing ambiguity, deception, and surprise. Finally, critical to proper learning is avoiding overestimating one's own strength, invulnerability, deterrence, or success potential.

To improve surmounting information interpretation barriers, machine-based big data analysis and artificial intelligence algorithms seem much more adept—and much more usefully incorporated—in addressing human cognitive frailty than security information unreliability. As noted previously, one of the major limitations of any form of computerized data analysis is an inability to correct for information that is rife with errors, distortions, and imprecision, some of which could be intentionally injected by adversaries. On the other hand, such dispassionate digital data processing can certainly reduce or eliminate the negative effects of human biases, preconceived notions, and selective attention. So despite the universal appeal of machine data processing—and specifically big data analysis—if facing overwhelming quantities of information, there should be a steadfast refusal to delegate all information interpretation to automated sources, even when they possess the most sophisticated artificial intelligence algorithms.

To improve detection of actual signals within a noisy information environment, it seems more important to spot rhetoric-reality discrepancies, especially between political statements and military actions, than to develop fixed predetermined abstract conceptual rules for determining data credibility and relevance. Specifically, "when the political conduct of the adversary is out of consonance with his military preparations, when he is talking softly but carrying a bigger and bigger stick, beware."[25] Identifying such discrepancies is not easy, for the crucial element of judgment enters the picture when attempting to distinguish between an insignificant and expected inconsistency and a highly significant and unexpected inconsistency. This warning about paying special attention to gaps and inconsistencies in adversaries' behavior seems particularly relevant to the foreign policy maker errors in failing to realize, despite a huge prior military buildup and mobilization, Saddam Hussein's intent to invade Kuwait in August 1990.

To improve recognition of disinformation and fake news, it seems more important to focus on a thorough investigation of the validity, reliability, and objectivity of data sources than on whether the information confirms or contradicts previous assessments. To discern an intelligence report's credibility, validity, and reliability and to minimize the chances of negative manipulation interference, it seems vital that "the collector provide as much information as possible on the origins of the report and the channels by which it was received, since the analyst who receives it will be almost completely dependent on such evaluations and comments in making his assessment."[26] When dealing with notoriously unreliable foreign security information, in the context of the digital data explosion this thorough vetting of data sources seems especially essential.

To improve assessment of manipulation outcomes, it seems more important to focus on whether enemies' attitudes and behavior transform than on whether one receives criticism or praise from friends. As noted in Chapter 4, in the case studies many applications of strategic ambiguity, deception, and surprise led to substantial questioning by allies or erosion of relationship with allies. However, the appropriate yardstick for successful manipulation ought to be the extent to which changes occur in targets, for if alliance foundations are deep and meaningful, friends should eventually be able to get over any adverse reactions they may have in this regard and accept the resulting changes in the security predicament. For example, following the 2016 Brexit vote, in the United Kingdom, Britain's allies in continental Europe were not likely to cut off UK relations as a result.

In a related manner, to improve manipulation initiators' consideration of long-run risks, it seems more important to focus on the dangers of local and regional instability than on the dangers of negative international community reactions. In many of the case studies, manipulation initiators were guilty of not looking ahead to consider the long-term consequences of their actions, and often local or regional dissatisfaction, tension, and instability ensued, whereas even severe outrage from the broader international community about such manipulative action rarely if ever had any meaningful impact. Although an ideal manipulation application generates no blowback, realistically one needs to pick and choose prudently where the blowback seems most likely to occur, and generally it seems better if it does so far away from the site of the application of strategic ambiguity, deception, or surprise.

To improve manipulation initiators' offensive thrusts, it seems more important to focus on increasing understanding of designated target vulnerabilities than on increasing acquisition of sophisticated manipulation tools. In 2001, the 9/11 terrorist attacks on the United States vividly highlighted the deception and surprise success frequently associated with carefully studying and exploiting target vulnerabilities:[27] "by exploiting their victims' vulnerabilities and lack of preparedness," al-Qaeda created a "greater potential to inflict high casualties and material damage" as well as the potential for being "highly demoralizing to the targeted state's general public and political leadership."[28] Similarly, during the 2008 Russian invasion of Georgia, "politically Russia identified and skillfully exploited the gap between Georgian and Western policies with respect to the 'frozen conflicts' of Abkhazia and South Ossetia."[29] Once an initiator becomes aware of these exploitable target vulnerabilities, it can maximize the range of strategic ambiguity, deception, and surprise options taken so as to overload a target's capacity to analyze, understand, and respond effectively to any one thrust. In contrast, in none of the cases were the basic manipulation methods or tools particularly novel or startling.

To improve manipulation targets' awareness of strategic ambiguity, deception, and surprise used against them, it seems more important to expect and analyze a broad interconnected pattern of continual manipulation efforts rather than to classify these as rare isolated incidents. Too often a potential manipulation target blithely assume that other countries, even adversaries, choose clear, predictable, and transparent policies toward it:

> We must be continually alert in such situations for the possibility of deception [along with other forms of foreign manipulation] and assume its likelihood— rather than its improbability. Rather than wax indignant over the enemy's "perfidy," as is our usual wont, we should be indignant at ourselves for failing to perceive in advance such a possibility. Bluntly, we need to be less trusting and more suspicious and realistic.[30]

In integrated manipulation assessment, target intelligence analysts need better balance, avoiding being too oversensitive or undersensitive in their manipulation reaction and in their thresholds for triggering alarm, too consistent or inconsistent in their warning, or too certain or tentative in their signal interpretation.

To improve manipulation targets' defensive responses, it seems more important to focus on promoting their vigilance and preparedness, heightening postmanipulation recovery and resiliency, than on reducing the frequency of strategic manipulation. Because of its ubiquitousness, strategic manipulation can come at a target from so many angles in so many forms that attempting to completely forestall it from occurring in the first place seems unrealistic in today's highly interpenetrated world. Unlike direct military attack, traditional deterrence approaches cannot work in preventing foreign ambiguity, deception, and surprise. As a result, attempting to optimize what happens following being victimized by manipulation appears to be a far more constructive defensive strategy, as minimal manipulation outcomes could indirectly discourage future foreign manipulation.

Improving Offensive Manipulation Under Information Overload

Given these management priorities, there appears to be significant room for improvement in strategic manipulation under information overload. Beginning with improvements in offensive strategic manipulation initiatives, several thrusts—shown in Figure 5.2—seem important. The primary challenge appears to be how to undertake enduring successful manipulation without risking the possibility of suffering overly harsh long-term security consequences.

Looking first at preparing prudently pre-manipulation, prior to undertaking strategic ambiguity, deception, and surprise, manipulation initiators should

PREPARING PRUDENTLY PRE-MANIPULATION

Planning
Engaging in more coherent advanced planning and identifying with unified commitment key strategic goals and methods for pursuing them
Intelligence
Collecting and analyzing better intelligence on relevant foreign targets' values, culture, expectations, and perceptual predispositions

CHOOSING FEASIBLE METHODS

Credibility
Using more realistic and fact-based manipulation to maximize credibility with targets, linking to their fears and identity aspirations
Confirmation
Operating more as much as possible within the sensitive confines of the preexisting beliefs of designated targets

INTENSIFYING OPERATIONAL TACT

Discretion
Ensuring more discretion, secrecy, and nontransparency among those in a position to know the true details of objectives and strategies
Channels
Searching more comprehensively to find or create appropriate information channels to influence designated targets

OVERCOMING TARGET RESISTANCE

Leverage
Taking better advantage of whatever leverage—such as that established through reputation, resources, or allies—is possessed at the time
Creativity
Promoting a combination of more inventive initiatives to overwhelm targets, avoiding repeated use of the same techniques over time

EVALUATING EQUITABLY POST-MANIPULATION

Feedback
Tracking better feedback on changes in targets' behavior, avoiding wishful thinking and adjusting nimbly to these changes
Contingencies
Calculating better feasible contingency paths out or escape routes if undertaken manipulation efforts fail or backfire

Figure 5.2. Improving offensive manipulation under information overload

have completed considerable background analysis on foreign targets. First, regarding planning needs, before manipulation begins, its most important objectives and means of attainment need to be clearly articulated.[31] These goals and strategies should be carefully considered in terms of relevance to national interests, feasibility, potential effectiveness, and perceived legitimacy. Use of ambiguity, deception, and surprise should not always be viewed as a weapon of the weak or as shameful, disgraceful, or immoral. Second, taking into account the need for intelligence, it appears crucial to understand the capabilities and weaknesses of an adversary target's intelligence surveillance and warning systems, including what kinds of data are tracked carefully versus downplayed or ignored, so that a manipulative initiator can carefully circumvent or disable them so as to avoid advance detection. Since no country's intelligence warning system can comprehensively cover all bases, understanding the target's intelligence prioritization system is vital. This intelligence should incorporate information about how targets would react and should behave and how the forms of manipulation chosen—strategic ambiguity, deception, and/or surprise—would induce that behavior. Most broadly, "deep intelligence penetration" about the target's predispositions is vital.[32]

Turning to choosing feasible methods, manipulation initiators should undertake approaches that, based on understanding a target's strategic culture and on becoming familiar with its perceptual predispositions, have the greatest chance of being plausible to foreign targets and of linking up to their ways of thinking.[33] Each manipulation technique must be "realistic and based on fact so it would be credible to the enemy," and "coordinated, integrated, cohesive, and accurate, without any gaps, to provide the enemy [the impression of] a continuous and linked chain of events."[34] Aiding this credibility quest would be a manipulation connecting directly to a target's deep-seated emotional fears and identity aspirations, such as those held by local nativist movements. Manipulation techniques should be used that reinforce distorted beliefs a target already possesses, with a high correspondence between projected images and "the preconceptions of the attacker,"[35] so as to take advantage of confirmation bias.

Moving to intensifying operational tact, manipulation initiators should be more sensitive in foreign use of ambiguity, deception, and surprise. Considering the need for discretion, better maintaining secrecy about intentions and capabilities is paramount.[36] One should "disguise operations by studying enemy vulnerabilities" to determine what areas to exploit; for example, during the 9/11 terrorist attacks on the United States, "bin Ladin's operatives had apparently relied on publicly available reports by the General Accounting Office to study the security weaknesses of sensitive U.S. facilities."[37] Distraction (refocusing attackers away from valued assets) and obfuscation (making a valued asset seem worthless) can be used to maximize target ambiguity, confusion, surprise, and unpreparedness about outcomes,[38] lowering chances of them launch any form of effective counterretaliation. Examining the importance of channels, it appears crucial not only to find and exploit existing formal and informal channels to reach a designated target but also to cleverly create new channels of influence—especially ones that create "more highly segregated systems of knowledge"[39]—if no satisfactory ones exist.[40]

As to overcoming target resistance, manipulation initiators should use strategies designed to overpower any foreign target counterretaliation. Regarding leverage, initiators need to carefully take stock of whatever special leverage they possess during manipulation over a target in areas where it is vulnerable and take advantage of it. Such leverage could be a function of alliances or interdependence relationships, a history of perceived obligations, skewed access to natural resources, or regional or global reputation. Regarding creativity, an initiator needs to be imaginative in expanding the range of manipulation techniques used at once against a target so as to overwhelm its capacity to interpret them correctly or to respond to them effectively.

Concluding with evaluating equitably post-manipulation, manipulation initiators should undertake a fair and balanced assessment of the kinds of emergency measures that should have been in place beforehand. Initiators should seek objective, balanced feedback about the success or failure of any just-undertaken manipulation so as to optimize future manipulation efforts. A key component of post-manipulation feedback is carefully assessing subsequent changes in a target's interpretation of events and in its behavior.[41] Special care needs to be applied by initiators not to become victims of wishful thinking by inadvertently downplaying any potential undesirable long-term aftershocks. In cases of manipulation failure, initiators should examine carefully the manipulation aftereffects to see exactly where and why the manipulation went wrong so as to be able to tweak its application to be more successful in future efforts. Regarding the importance of contingencies, when manipulation is disastrous, it is important to develop ahead of time multiple alternative courses of action that can minimize long-term damage.

Improving Defensive Response Under Information Overload

Considering improvements in defensive strategic manipulation response under information overload, shown in Figure 5.3, several thrusts seem vital. The primary defensive challenge is formulating responses likely to improve ongo-

---

INTERPRETING DATA CAUTIOUSLY

Wariness
Being more wary about data falling neatly into a single pattern, ignoring other reasonable meanings, and being more vigilant about distorted data
Realism
Shunning more tendencies to overestimate one's own invulnerability and underestimate undesired intrusions or outcomes

---

DEEPENING PREDICAMENT UNDERSTANDING

Sensitivity
Sharpening sensitivity better to other cultures and refraining from projecting one's own values or modes of reasoning onto others
Respect
Avoiding more seeing adversaries as mindless and irrational and underestimating their sophistication

---

REFINING THREAT ASSESSMENT

Training
Enhancing better training of intelligence analysts to recognize their own biases and to accept uncertainty in data and predictions
Advice
Having security policy makers seek and receive better advice and assistance from capable, prudent, and experienced outside parties

---

KEEPING ADVERSARIES OFF-GUARD

Confounding
Maximizing initiators' uncertainty about impacts and reactions, distracting them from valued assets and making gains seem worthless
Fluidity
Building a more multifaceted toolkit of fluid responses, openness to the unexpected, and eagerness to adapt

---

ENHANCING RECOVERY CHANCES

Learning
Refining learning better from past defensive successes and failures, with regular independent internal and external reviews
Resiliency
Making everyone within one's society more vigilant, prepared, adaptive, recuperation-oriented, and resilient when facing unexpected change

---

Figure 5.3. Improving defensive response under information overload

ing data interpretation and to thwart not only past but also future foreign manipulation. The underlying thrust revolves around trying to switch information overload's impact from enhancing to constraining opportunities for cross-national manipulation.

Looking first at interpreting data cautiously, manipulation targets should be more careful in how they draw meaning from incoming foreign information. As to the need for caution, often information consumers interpret communication on its face value, especially if it is "comforting, convincing, or highly classified."[42] Instead, it appears especially important to "be wary of information which falls too neatly into a single pattern that seems to exclude other, no less reasonable possible courses of action."[43] "Always conduct a reality check from not only your own perspective but also that of the opponent"—"if everything is crystal clear and consistent with your best-case scenario, and the adversary behaves just like you would in similar circumstances, you are probably being deceived."[44] Moreover, better safeguards should be instituted at all societal levels to ensure the credibility and accuracy of incoming foreign security data. Maintaining constant circumspection about what incoming signals mean and what will work best defensively is critical. Considering the need for realism, manipulation targets should become more attentive to possibilities of dire consequences from foreign manipulation and more determined to avoid complacency about the adequacy of existing defenses and countermeasures. Rather than emphasizing existing immunity to disruption, both government officials and private citizens need to become more familiar with the methods and practices of manipulation[45] and potential negative security consequences.

Moving to deepening predicament understanding, manipulation targets should find better ways to develop a probing operational understanding of foreign manipulation initiators. Regarding the need for sensitivity, targets need to improve decoding of preconceptions, standard operating procedures, and underlying values of initiator cultures. Particularly for Western targets, prone to project their ways of thinking onto initiators, there should be hypersensitivity to vast cultural and interpretive divides. Regarding respect, targets should refrain from viewing potential initiators as so backward or irrational that they could not possibly orchestrate successful foreign manipulation. Without more dispassionate and open-minded assessment of adversary capabilities and intentions, targets seem likely to inappropriately dismiss foreign manipulation dangers, exemplified by pervasive demeaning views of terrorists:

> We are . . . inclined to see terrorists as fiends, wild-eyed expressions of evil, diabolical but two-dimensional, somehow alien—in a word, inhuman. Government officials routinely denounce terrorists as mindless fanatics, savage barbarians, or, more recently, "evil-doers"—words that dismiss any intellectual

content. The angry rhetoric may resonate with apprehensive homeland audiences, but it impedes efforts to understand the enemy. We cannot formulate multidimensional responses to terrorism that combine physical destruction with political warfare if we do not see our adversaries as anything other than comic-book villains.[46]

In contrast, adversaries of all stripes often possess well-thought-out manipulation strategies that can be quite effective for their nefarious manipulative ends.

Turning to refining threat assessment, manipulation targets should protect against the distorting effects of their own preconceptions and be open to outside assistance in interpreting and responding to foreign dangers. There needs to be much more extensive and focused training for government intelligence collectors and analysts about contaminating influences surrounding interpreting strategic ambiguity, deception, and surprise.[47] Such training, which needs regular updating, should include improving government officials' awareness and correction of distortions introduced by their own biases and by existing bureaucratic inertia. More exposure to information-processing limitations associated with human cognitive frailty and organizational decision inflexibility may be quite helpful in this regard. There also needs to be much greater openness—particularly among policy makers in countries without substantial intelligence capabilities—to request and receive outside help in interpreting and responding to foreign manipulation. There should be no perceived internal or external shame, humiliation, or revelation of weakness in requesting this kind of aid. Given the poor signal-to-noise ratio and the subtlety of signals sent across national boundaries, even targets with significant intelligence prowess may benefit from this kind of assistance.

Pondering keeping adversaries off-guard, a manipulation target should find ways to prevent foes from finding its manipulation responses too predictable. Manipulation targets should take steps to conceal the security impacts of foreign manipulation and the nature and success of their own responses so as to keep adversaries confused. The more uncertain aggressors are about the cost-benefit ratio of their past manipulation efforts, the more hesitant they will be to undertake manipulation against the same targets in the future. The more flexibility and adaptability a manipulation target displays, the more confounded a manipulation initiator will be about what will work against it. Because foreign manipulation is constantly in flux, it is difficult for either government officials or private citizens to keep abreast of all of its possible varieties, particularly given fixed standard operating procedures and onerous bureaucratic inertia. Avoiding paralysis in target security policy responses is crucial. In many cases, targets of high-stakes foreign manipulation would be wise to set up special emergency task force arrangements, allowing those responsible to cut through the red tape and act quickly and decisively in manipulation responses.

Concluding with enhancing recovery chances, foreign manipulation targets should find better ways to learn from their past defensive mistakes and to be more resilient in the face of foreign intrusions. Targets need to derive better lessons for future preparation and response from past foreign victimization, avoiding a common dysfunctional tendency to use selective attention to avoid spotting glaring defensive weaknesses. Bringing in fresh eyes to help interpret and respond to outside manipulation seems highly advisable, as often consultation is inadequate with those "who might provide fresh insights and different perspectives."[48] Independent internal and external reviews of manipulation responses and their impacts need to occur regularly to forestall societal complacency. Second, regarding resiliency, targets need to prepare defensively for foreign manipulation by increasing post-manipulation resiliency capabilities—the ability to quickly rebuild and restore damaged target infrastructure and authority credibility. Manipulation targets should promote alertness by everyone about the dynamics of foreign manipulation and potentially successful responses, stepping back and undertaking a probing continuous examination of changing dangers posed by its foes and the possible strategies they could use to pursue them. Greater target awareness of manipulation modes, including awareness by the mass public and the media, seems pivotal to successful defense.[49] This vigilance should include attentiveness to warnings about low-probability, high-impact threats—"it is not the odds in themselves which determine the importance of the warning judgment to the policymaker, but the potential consequences of the action if it should occur."[50] In international security affairs, there should be a continuous expectation of ambiguity, deception, and surprise in global communication.

Addressing Specifically Strategic Ambiguity, Deception, and Surprise

The suggested general improvements in offensive manipulation initiatives and defensive manipulation responses highlight the broad changes useful to manage strategic manipulation under information overload. However, some more specific needed advances—summarized in Figure 5.4—apply specifically to strategic ambiguity, deception or surprise.

IMPROVING STRATEGIC AMBIGUITY MANAGEMENT UNDER INFORMATION OVERLOAD

Offensive
Becoming able to predict more confidently because of specialized intelligence which interpretations targets will choose of ambiguous signals
Taking better advantage of the resulting flexibility (keeping options open) and plausible deniability (eluding undesired accountability)
Defensive
Performing more balanced response analyses, not overreacting to minor threats or underreacting to major threats
Figuring out how to translate ambiguity into opportunities for better problem solving by creatively questioning underlying assumptions

---

IMPROVING STRATEGIC DECEPTION MANAGEMENT UNDER INFORMATION OVERLOAD

Offensive
Analyzing better what kinds of trickery would best exploit targets' intelligence holes and trusted sources
Selecting better when to use deception given its operational limits—its occasional unavailability and its difficulties of repeated usage
Defensive
Having those possessing effective counterdeception measures better assist allies lacking such tools to hide or manage vulnerabilities
Publicizing known manipulation initiators' deception uses so as to expose and potentially humiliate them globally

---

IMPROVING STRATEGIC SURPRISE MANAGEMENT UNDER INFORMATION OVERLOAD

Offensive
Becoming more familiar with the set of target expectations relevant to the proposed surprise action
Understanding better target intelligence surveillance and warning system weaknesses so they can be disabled, corrupted, or circumvented
Defensive
Thinking more creatively outside the box, deviating from typical planning to anticipate, react to, and recover from surprise moves
Working in advance through beefing up strategic warning to better contain negative security consequences emanating from surprise

---

Figure 5.4. Addressing specifically strategic ambiguity, deception, and surprise

Improving Strategic Ambiguity Management Under Information Overload

For strategic ambiguity, the key offensive improvements are (1) becoming able to predict more confidently because of specialized intelligence which interpretations targets will choose of ambiguous signals and (2) taking better advantage of the resulting flexibility, keeping options open, and of plausible deniability eluding undesired accountability. The key defensive improvements are (1) performing more balanced response analyses, not overreacting to minor threats or underreacting to major threats, and (2) figuring out how to translate ambiguity into opportunities for better problem solving by creatively questioning underlying assumptions.

Ambiguity surrounding important decisions is decidedly uncomfortable; in recent stressful foreign policy crises, it has been a consistent root of major foreign misperceptions[51] and has consequently led to many catastrophic blunders (such as the results of the ambiguity surrounding the presence of WMD in Iraq before the 2003 American invasion). Typically "we often handle ambiguity poorly and we can do better"; while the world has "always involved handling ambiguous information in high-stakes circumstances," today to many manipulation targets "the world feels more overwhelming and chaotic than ever."[52]

In light of these management challenges, the most essential step for manipulation targets to manage strategic ambiguity better is a dramatic refocus in attitude, emphasizing receptivity to and respect for ambiguity as a positive opportunity rather than as a disruptive constraint, generating frustration about ensuing interpretation and decision dilemmas. After all, ambiguity is what makes problems so interesting to intelligence analysts.[53] More specifically, embracing ambiguity and the subsequent interpretation struggle can be an "emotional amplifier" enhancing creativity, improving problem solving, increasing justified questioning of black-and-white analyses, and opening up divisive controversies for discussion—"ambiguity can help us learn something new, solve a hard problem, or see the world from another perspective."[54] To improve our thinking in such a situation, we need "to learn to adjust agreeably to new circumstances."[55] In this vein, sometimes pressure to reduce ambiguity can be disastrous, for it is quite common to experience "ambiguity-reducing deceptions" that make targets "quite certain, very decisive, and wrong."[56] Overall, within "an increasingly complex, unpredictable world," what appears to matter most for successful survival is "how we deal with what we don't understand."[57]

When focusing on foreign security information and communication, admittedly finding appropriate ways to manage strategic ambiguity seems especially challenging, because of the major impediment of security data unreliability interfering with the use of confirming and disconfirming evidence as a means of discriminating between conflicting interpretations so as to make better decisions. However, even here refocusing one's orientation toward strategic ambiguity may help improve its security effects. Manipulation initiators could emphasize more taking advantage of the resulting greater flexibility provided in keeping several paths open, ability to reach international security agreements whose vague language allows all signees to interpret their commitments differently, capacity to stall and defer difficult decisions, and plausible deniability internally and externally surrounding accountability for unpopular decisions and outcomes.

Improving Strategic Deception Management Under Information Overload

For strategic deception, the primary offensive improvements are (1) analyzing better what kinds of trickery would best exploit targets' intelligence holes and trusted sources and (2) selecting better when to use deception given its operational limits—its occasional unavailability and difficulties of repeated usage. The primary defensive improvements are (1) having those who possess effective counterdeception measures better assist allies who lack such tools in hiding or managing vulnerabilities and (2) publicizing known manipulation initiators' deception uses so as to expose and potentially humiliate them globally.

For offensive initiation, sometimes deception is not an available option, as concealment may be impossible, and deception may work only once, after which the deceived party figures things out. Nonetheless, improving offensive deception management can be aided by learning more about what a target relies on as the most credible intelligence sources; for example, al-Qaeda probed what telephone lines were being monitored by enemy governments to spread misleading information.[58] Notably, the careful intelligence gathered by the terrorists masterminding the 9/11 terrorist attacks served as a model of offensive deception management; "familiarity with the West" was "one of the characteristics most sought after by the Al Qaeda leadership in their selection of the pilots and cell leaders," enabling "the hijackers to integrate more easily into U.S. society, thereby fulfilling the crucial requirement of remaining anonymous and indistinct."[59]

For defensive responses, typically targets have been duped, at least temporarily, by adversaries' deception, leading to gross and costly miscalculations. Thanks to the pervasiveness of strategic deception, targets should consider more carefully correcting estimates of adversaries' force levels so as to incorporate possibilities of deception.[60] A sound defensive response to deception is to widely publicize previously detected deception uses to expose and potentially humiliate globally the manipulation initiators. Transnational activist groups promoting transparency and seeking to expose corruption, as well as Internet-based social media sites promoting truth-telling and seeking to expose fraud, provide excellent eager and willing outlets for facilitating this kind of global awareness of strategic deception that can sometimes help deter future manipulation initiatives.

As a common deception target, the United States really needs defensive improvement:

> The United States' view of the world is not conducive to understanding fanatical zealotry, systematic secrecy, sustained sacrifice without the prospect of success, the paradox of defeat as victory, and the determination to deny and deceive. This same intellectual outlook and psychological predisposition make it difficult for U.S. officials to establish and operate a comprehensive, coordinated counterdeception program.[61]

Overcoming Western/non-Western cultural barriers, perhaps with more sharing of counterdeception measures among allies to mask or manage key vulnerabilities, seems crucial to success in this regard.

When dealing specifically with foreign security information and communication, strategic deception seems especially hard to detect because of the poor signal-to-noise ratio characteristic of today's defense information overload. Given the high levels of secrecy and security classification associated with sensitive defense data, intelligence agencies within targets are often thwarted because offense seems so far ahead of defense: standard means of intelligence collection can be stymied by encryption and limited information dissemination,[62] and visible protective monitoring can show deceivers right where to channel credible false signals. However, even here more concerted target efforts could help conceal the true scope and nature of key intelligence sources exploitable through deception.

Improving Strategic Surprise Management Under Information Overload

For strategic surprise, the primary offensive improvements are (1) becoming more familiar with the set of target expectations relevant to the proposed surprise action and (2) understanding better target intelligence surveillance and warning system weaknesses so they can be disabled, corrupted, or circumvented. The primary defensive improvements are (1) thinking more creatively both outside the box, deviating from typical strategic planning, to anticipate, react to, and recover from surprise moves and (2) working in advance through beefing up strategic warning to better contain the negative state and human security consequences emanating from strategic surprise.

There are several reasons why it is so difficult for targets to take effective preemptive action to prevent strategic surprise and why fewer defensive moves seem available to forestall surprise than to forestall ambiguity or deception:

Hedging against surprise is hard because the causes are deeply embedded in human psychology, political uncertainty, military complexity, and organizational viscosity. Making surprise less consequential is difficult because . . . it can be done in only two ways, neither of which appeals to Western societies. One is by gross overinsurance in force levels, developing capabilities so superior that they can triumph even if a large proportion is neutralized at the outset. . . . The second logical solution is scarcely more palatable to Western political leaders: the development of such high levels of peacetime readiness, unconstrained by a requirement for collective political authorization to change posture, that negligible warning or decision time are necessary to prepare for combat.[63]

Such challenges explain why deception is "almost always successful in achieving surprise, regardless of the experience of the adversary."[64] Typically, even targets with excellent surveillance capabilities have been caught unaware by strategic surprise, and the short-run security consequences have often been devastating. Success seems especially likely when surprise initiators understand a target's expectations and its intelligence system weaknesses.

Nonetheless, for surprise manipulation targets there are available defensive response options, which even if not always able to prevent surprise can reduce its negative effects or improve preparedness and resiliency. Thinking creatively outside the box about what could happen and about how to respond seems exceptionally important, and here intelligence analysts need to overcome a common tendency to "be risk averse and more concerned with avoiding mistakes than with imagining surprises."[65] One means for intelligence analysts to avoid getting in a rut when trying to imagine unexpected events is to examine in a special way past strategic surprises:

Analysts should keep a record of unexpected events and think hard about what they might mean, not disregard them or explain them away. It is important to consider whether these surprises, however small, are consistent with some alternative hypothesis. One unexpected event may be easy to disregard, but a pattern of surprises may be the first clue that your understanding of what is happening requires some adjustment, is at best incomplete, and may be quite wrong.[66]

If targets are to recover from strategic surprise, thanks to advanced preparation, they "must be robust and resilient enough to bounce back from crises and prepared enough to minimize the damage when devastating events occur."[67] Given that "intelligence failures are inevitable, we must learn to live with ambiguity—the best way of avoiding surprise is being able to cope with it, once it has taken place."[68] In addition, targets need to work in advance to try to contain as much as possible the state and human security consequences;[69] such containment emphasizes reducing the severity of short-term security disruptions by enhancing target warning systems, squeezing in a little more time to prepare for aftershocks and to minimize the chances of direct human carnage and property damage.[70] The hope here is to "adopt defenses to deflect the most dangerous possibilities or at least trigger an early warning."[71] Beefing up strategic warning—through clarifying the warning mission, increasing relevant resources, and improving estimative analysis[72]—constitutes an important component of any successful defensive response to strategic surprise.

When dealing specifically with foreign security information and communication, strategic surprise seems easiest to spot when facing unexpected conventional military attacks in which significant visible movement necessarily occurs beforehand. Yet in this case "even with the best equipment in the world, militaries around the world frequently have been surprised by their adversaries."[73] A common explanatory obstacle is not the unavailability of defensive response options to cope with strategic surprise but rather the emergence of target government complacency after a long period of calm, causing existing response options not to be properly applied[74]—"the 'it can't happen here' syndrome is a major obstacle to preventing, learning from, and properly preparing for crises."[75] When target government foreign security policy makers are surprised, their typical knee-jerk responses are usually ineffective: assigning blame so as to take corrective action is difficult because the responsibilities are normally so diffusely scattered;[76] "wounded and humiliated, governments subjected to such surprise attacks are more likely to opt for harsh and risky responses against the perpetrators, thereby running the risk of drawing the two sides into an escalating, often protracted confrontation that is costly in both human and economic terms."[77] To solve the problem of blame diffusion, and to overcome normal government bureaucratic organizational decision inflexibility, there should be "better interagency coordination and cooperation," reducing immediate vulnerability to attack and increasing timely capacity to respond,[78] and then directly circumventing the barriers to information sharing posed by excessive compartmentalization and application of the need-to-know principle.[79] To overcome the problem of quick impulsive harsh target responses, reflective time needs to be available for "freewheeling, unconstrained, uncritical" consideration of response ideas, entailing "deferred judgment" before policy decisions are reached.[80]

Conclusion

THE SEARCH FOR MEANING is a universal human pastime. Although that quest is often quixotic, doing so provides a central purpose for existence. We feel a pressing need to understand the world around us and to find a way to interpret the surrounding chaos. We reject the idea of living in a world of illusion shrouded by secrecy and mystery, desperately wanting to find out what is really going on so that we can take steps to maintain what works and change what fails. Discovering a way to interpret information accurately is central to making decent predictions, performing cost-benefit analyses, assessing risk, reaching sound decisions, and ultimately taking constructive action.

When information overload and strategic manipulation swirl around us, our search for meaning is frequently stymied. As a result, both government officials and private citizens can become confused, fearful, lost, paralyzed, angry, frustrated, and resentful, seeing no clear pathway to escape from this predicament. They feel marginalized by trends they cannot control, unable to find a secure place for themselves within the rapidly changing system, and highly uncertain about the future. Although they spend record amounts of time and effort assimilating information and receiving and sending messages, with all the data they could ever want at their fingertips, they do not know what to believe or think, and many feel overwhelmed by the cost and complexity of what is required to get a handle on what is really transpiring. Citizens wonder whether national governments or other authority structures have significant better plans to remedy this situation, and national governments wonder whether they can come up with better ways to enhance the search for meaning and to protect citizens from the insecurity-enhancing ravages of foreign manipulation.

Rising Informal Influence in International Relations

The wide use of strategic manipulation under information overload is one symptom in today's global security setting of how much subtle, nuanced, indirect, intangible means of cross-national influence and control have replaced crude territorial takeover. Indeed, "perhaps for the first time in history, the ability to inflict damage and cause strategic dislocation is no longer proportional to capital investment, superior motivation and training, or technological prowess."[1] Instead, "informal penetration"[2] is the new norm, requiring a more comprehensive understanding of adversaries' formal and informal—and top-down and bottom-up—planning and decision-making processes.[3] Many crucial international messages are now transmitted through informal channels. Regardless of the success rate, the desire to manipulate information—figuring out for oneself what is really going on while shaping others' perceptions—has increased. Although information overload can mute its impact, propaganda is experiencing a rebirth in an era in which "fake news" and dubious global trends increasingly dominate the headlines and in which all are more uncertain than ever about their security.

Within the anarchic global security setting, selective use of strategic ambiguity, deception, and surprise can sometimes be a cheap and effective way to pursue foreign security policy ends and to get one's message heard effectively. For offensive initiators, manipulation through informal channels seems especially successful when confirming deeply seated emotional fears and identity aspirations, blaming problems on foreigners and distracting attention away from domestic failings, or portraying aggressive actions as defensive within a sphere of influence. For defensive targets, manipulation responses through informal channels seem especially successful when emanating from vigilant, prepared, adaptive, and resilient societies unified in their resolve.

Strategic Manipulation's Ethical Concerns

Even if one steadfastly contends on a moral basis that strategic manipulation's short-term initiator benefits are inescapably outweighed by significant long-term costs to international system stability, today's operational global norms tilt toward tacitly condoning the use of strategic manipulation against foreign targets. Most countries' political leaders use strategic ambiguity, deception, and surprise, targeting mostly foreign adversaries but also at least occasionally allies or even domestic private citizens. For example, strategic deception of adversaries is generally considered to be normatively tolerable and indeed is a regular component of diplomacy[4]—"the use of deception has been tacitly accepted as quasi-legitimate in relations between states, and states have learned to expect it as a way of life."[5] Regardless of its security consequences, this manipulation usually occurs with impunity, with sufficient cynicism existing among victims of strategic ambiguity, deception, and surprise that intrusions do not usually translate into meaningful countermeasures—punishment or reprisals—afterward. Within the current anarchic world, foreign manipulation flourishes because of both (1) the inherent credibility confusion among targets and outside observers and (2) the differing values prevailing about the legitimacy of various types of manipulation.

Despite this tacit global acceptance, ethical concerns surround strategic manipulation. Trust is a truly precious and fragile commodity in international relations, and information overload can cloud and conceal its foundation while strategic ambiguity, deception, and surprise constantly eat away at its meaning and operational impact. The pervasive absence of a base level of trust and respect among states and nonstate groups can easily destroy any cornerstone for constructive collective action and facilitate global anarchic violence.

Internal or external strategic manipulation by those publicly espousing virtues of enlightened democratic transparency highlights a kind of morally questionable hypocrisy that serves nobody's interests. All manipulation is not equally ethically problematic: offensive use of foreign manipulation, interfering with the principle of self-determination of peoples, seems more controversial than defensive obfuscation responding to foreign manipulation; targeting one's own citizens seems more controversial than targeting foreign adversaries, because it violates citizens' "right to know"; and maliciously falsified data seems more controversial than benignly fabricated data because lacking resources to provide accurate information is generally considered both understandable and excusable.

Facing these significant global ethical concerns, it seems foolish to avoid confronting them directly. No matter how challenging growing information access and international strategic manipulation may be, neither the "ignorance is bliss" response nor the "delegate understanding and judgment to experts" response seems to be a viable remedy.

Because information overload has expanded both the frequency of and implicit tolerance for strategic manipulation in all spheres, if a state wished to avoid ethical tainting by having its communication labeled as manipulative by the international community, its words would need to be clear, accurate, and predictable, backed up with tangible supportive actions enhancing their credibility. It appears as if the only way to have this kind of "honest broker" image in a disingenuous world is to marshal multiple forms of confirming evidence to prove it. However, because of entrenched global skepticism, shunning strategic manipulation completely and attempting to communicate with others truthfully and transparently on all occasions seems unlikely to achieve desired confident predictable cross-unit cooperation and can easily lead to security policy failure.

Information Transparency Paradox

Given these ethical considerations, promoting transparency, accuracy, precise clarity, and predictability in communication is not always best for global security, particularly when dealing with ruthless adversaries. On the surface, this kind of communication may seem to be the most virtuous and laudable, with many citizens preferring predictable transparent clarity over a world where subtle signaling, nuanced communication, and devious covert intentions run rampant. However, within a global setting full of ambiguous, deceptive, and surprising signals, transparency may be functionally indistinguishable from manipulation, and sometimes misinterpretation, confusion, or illusion is actually desirable for international stability. In a world where distortion has become the norm, operating as if direct honesty in global communication will shine through, reaping rewards of belief, friendship, cooperation, and compliance, can be naïve.

Paradoxically, particularly under information overload, increased transparency can often deepen international conflict. Because "transparency affects the *amount* and *type* of information observers receive rather than its interpretation," it may unintentionally "exacerbate conflict by providing too much information and making it more difficult to reach accurate judgments about another state's intentions"—"the sheer volume of information that transparency provides may not only create confusion and ambiguity in international conflicts, but may also undermine behind-the-scenes efforts at conflict resolution."[6] This unintended side effect is hard to forestall.

High levels of transparency do not guarantee either accurate interpretation of information, balanced awareness of contending views, or appropriate decisions. Within the overwhelming flood of information, the contemporary world is giving us more reality and more truth than we can comfortably handle, and because of their sensational appeal, more aggressive, belligerent, and extremist views may receive more media and mass public attention than modest, subtle, and conciliatory gestures.[7] For this reason, complete transparency about dire national threats or vulnerabilities could lead to widespread public fear, unrest, and dissatisfaction with government, perceiving it as incompetent because it cannot readily contain or eradicate alleged imminent dangers. In such cases, political leaders become reluctant to publicly communicate details or to let citizens transparently know the full range of dismal contingencies:

> Suppose the unthinkable happened, and terrorists struck New York or another big city with an atom bomb. What should people there do? The government has a surprising new message: Do not flee. Get inside any stable building and don't come out till officials say it's safe.
>
> The advice is based on recent scientific analyses showing that a nuclear attack is much more survivable if you immediately shield yourself from the lethal radiation that follows a blast, a simple tactic seen as saving hundreds of thousands of lives. Even staying in a car, the studies show, would reduce casualties by more than 50 percent; hunkering down in a basement would be better by far.
>
> But a problem for the Obama administration is how to spread the word without seeming alarmist about a subject that few politicians care to consider, let alone discuss. So officials are proceeding gingerly in a campaign to educate the public.
>
> "We have to get past the mental block that says it's too terrible to think about," W. Craig Fugate, administrator of the Federal Emergency Management Agency, said in an interview. "We have to be ready to deal with it" and help people learn how to "best protect themselves."[8]

Thus data transparency seems safest when news is comforting, or at least not totally alarming.

One data transparency commonality across societies is that all pay more lip service to its pursuit than actually taking concrete steps to ensure its full implementation. Mass public resentment of this pattern is highest within democracies. Given the wide variety of information sources available today, it is not even clear whether dramatically increased government security transparency would make a huge difference, for much new information would likely get lost in the flood of non-state-generated data. So with growing information overload, continuing overtransparency dangers, and uncertain transparency payoffs, the status quo of considerable nontransparency seems likely to persist for the foreseeable future.

Future Global Data Shock Dangers

With global information overload, security data unreliability, human cognitive frailty, organizational decision inflexibility, cultural diversity, and political anarchy enhancing foreign signal misinterpretation and manipulation, an ominous Orwellian future may emerge, in which communication controllers spin data in whatever direction they wish. Without foolproof fact-checking, everyone could be at the mercy of skilled manipulators. Two extreme reactions seem possible, both undermining the original goals of having the world wired—to prevent people from being duped by one-sided propaganda, and to open up policy decisions to bottom-up influences.

The first position is a Luddite anti-data rebellion. Because of data complexity and confusion, people could throw in the sponge, becoming fed up with information overload and desensitized to the value of data, and tacitly renounce their right to perform their own data interpretation. Rejected also would be the value of evidence-based decision making based on scientific empirical analysis, replaced by intuition or seat-of-the-pants wisdom. If this depressing pattern ensues, it could threaten the integrity of democratic processes because of citizens opting out of the political process or lacking the tools or the inclination to challenge a political regime's policies.

The second position is blind faith in the veracity of computerized data analysis. Because information overload seems to give the advantage to machine over man in data interpretation, technology advocates may push dispassionate computerized data assessment to totally replace defective human judgment as the basis for foreign security policy. Cutting humans out of the loop would seem to promote both comprehensiveness and objectivity. Dazzled by computers' technological capabilities, people could place too much trust in their value, subscribing to a kind of "data fundamentalism" contending that numbers cannot lie and constitute objective truth.[9]

The Need for Collaboration in Confronting Global Data Shock

Given rapidly changing conflicting data, intelligence analysts need to reject the notion that obtaining extra information guarantees a true and up-to-date threat picture of adversary intentions or actions, instead accepting more inconclusive, nuanced, tentative, and qualified judgments: "No matter how good, fast, or innovative technology becomes, people will never have 'enough' intelligence" to make decisive assessments and predictions.[10] Pressing for definitive conclusions unjustified by evidence can corrupt analysis and be just as likely as data misinterpretation to produce negative security consequences. Thus perhaps the optimal operational success metric in intelligence collection should simply be "an edge over the other guy"—better intelligence than that of the adversary.[11] Toward this end, multiple forms of intensified collaboration are crucial.

The Need for Responsible Human-Machine Collaboration

Intelligence analysts decidedly need extra help, for "perhaps in no other industry is the reliance on large and dynamic data streams more apparent than intelligence analysis" of foreign security threats,[12] and the data revolution has outstripped "the capabilities of intelligence organizations to analyze and interpret information."[13] Indeed, experienced well-trained professionals can easily fall prey to sophisticated foreign disinformation. Alone neither humans nor automated artificial intelligence systems can cope with global data shock: given low data quality, handing control off to even a high-quality black-box autonomous system is problematic, for it can perpetuate misinterpretation and inappropriately reinforce benign fabrication and malicious falsification; yet given high data quantity, allowing decision-making power to rest solely with humans is equally troublesome, because of their myriad limitations linked to cognitive frailty, organizational inflexibility, and ultimately vulnerability to outside strategic manipulation.

The solution is to have automated data-driven decision systems and intelligence analysts work together, so that computers assist rather than replace humans. In this effort, data science needs to venture well beyond simple big data analysis information mining: given high-volume data, instead of past rigid rule-based artificial intelligence (AI) systems, modern data science would employ reinforcement learning techniques to swiftly and automatically determine which decisions to make; but given the pervasiveness of low-quality security data, humans would need to collaborate closely to prevent unreasonable or misdirected outcomes. To optimize such collaboration, more research is needed to leverage the distinctive strengths of humans and computers in making difficult data-driven decisions quickly, safely, effectively, and legitimately; address the novel challenges arising inevitably from starting to deploy automated systems in real-world security intelligence settings; and understand the political ramifications of allowing AI systems to identify and manage emerging cultural issues.

The Need for Responsible State-to-State Collaboration

International inequality in the capacity to cope with global data shock is growing, and so increased cross-state cooperation—by which those with expertise help those without—seems essential, because the ripple effects of misinterpretation and manipulation rarely stay within a single country. Regarding data access, in several places in the world (even within the United States) there is no reliable Internet availability, handicapping basic predictable access to vital data. Regarding technical skills to manage incoming data, inequalities seem even greater. If those with data access and expertise continue to operate out of narrow self-interest and ignore friends needing help, debilitating gaps will widen between the information "haves" and "have-nots." For meaningful cross-national collaboration, increases in signal credibility and trust are prerequisites, along with improved intelligence analyst training about when and how best to share insights with outsiders.

The Need for Responsible Citizen-Government Collaboration

Better collaboration is also essential between citizens and their national governments in overcoming information interpretation barriers, deriving meaning from signals, resisting outside manipulation, and promoting state and human security. Like state-to-state collaboration, citizen-government collaboration necessitates first developing increased mutual trust. This can be accomplished through (1) a more open and honest two-way discussion about common confusions and vulnerabilities when facing information overload and strategic manipulation and (2) a more mutually acceptable balance about sacrifices in personal freedoms and privacy required in exchange for better foreign threat monitoring and prevention.

Concluding Thoughts

Ultimately, battling dysfunction in a world full of information overload and signal manipulation is a burden that falls on all of us, not just on national governments, private businesses, information technology experts, and international organizations. Compared to the past, "the fundamental change is that more of the responsibility for knowing what is true and what is not now rests with each of us as individuals," entailing new obligations and risks.[14] While most of us lack the time, energy, expertise, or motivation to reflect on how best to deal with information overload and strategic manipulation, and such efforts might well entail considerable emotional stress, navigating strategic manipulation under information overload has become a truly universally vital survival skill. If the new international norm involves being continuously overwhelmed with ambiguous, deceptive, and surprising data, then the least we can do is to try to have the ensuing offensive and defensive battle between initiators and targets be more prudently constrained. Because the security stakes are so high, we urgently need significant progress in confronting global data shock.

Notes

Introduction

1. For an early discussion of the ramifications of the signal-to-noise ratio in information interpretation, see Roberta Wohlstetter, *Pearl Harbor: Warning and Decision* (Stanford, CA: Stanford University Press, 1962), 392.

Chapter 1

1. Alvin Toffler, *Future Shock* (New York: Random House, 1970); "Too Much Information: How to Cope with Data Overload," *The Economist*, June 30, 2011, http://www.economist.com/node/18895468.

2. Yaacov Y. I. Vertzberger, *The World in Their Minds: Information Processing, Cognition, and Perception in Foreign Policy Decisionmaking* (Stanford, CA: Stanford University Press, 1990), 87.

3. Bree Nordenson, "Overload! Journalism's Battle for Relevance in an Age of Too Much Information," *Columbia Journalism Review*, November–December 2008, http://archives.cjr.org/feature/overload_1.php.

4. "Fake News," *60 Minutes*, March 26, 2017, https://www.cbsnews.com/video/the-attack-in-garland-fake-news-chess-country.

5. "Too Much Information."

6. Daniel J. Levitin, *The Organized Mind: Thinking Straight in the Age of Information Overload* (New York: Penguin, 2014), 6.

7. William J. Lynn III, "Defending a New Domain: The Pentagon's Cyberstrategy," *Foreign Affairs* 89 (September–October 2010): 98.

8. John Arquilla and David Ronfeldt, "Cyberwar Is Coming!" in *In Athena's Camp: Preparing for Conflict in the Information Age*, ed. John Arquilla and David Ronfeldt (Santa Monica, CA: Rand, 1997), 23. See also "Fierce Cyber War Predicted,"

CSO, March 3, 2003, https://www.csoonline.com/article/2115962/data-protection/fierce-cyberwar-predicted--post-9-11-training-sped-rescue-at-station-fire--your-home.html.

9. Martin C. Libicki, "Information War, Information Peace," *Journal of International Affairs* 51 (Spring 1998): 411–412.

10. Joe Havely, "Why States Go to Cyber-War," *BBC News*, February 16, 2000, http://news.bbc.co.uk/1/hi/sci/tech/642867.stm.

11. Bernard I. Finel and Kristin M. Lord, "The Surprising Logic of Transparency," *International Studies Quarterly* 43 (June 1999): 315.

12. Ibid., 315.

13. Executive Office of the President, *Big Data: A Report on Algorithmic Systems, Opportunity, and Civil Rights* (Washington, DC: U.S. Government Printing Office, 2016), 21.

14. Finel and Lord, "Surprising Logic of Transparency," 319.

15. Nicholas Carr, *The Shallows: What the Internet Is Doing to Our Brains* (New York: Norton, 2010), 2–3.

16. David Shenk, *Data Smog: Surviving the Information Glut* (New York: HarperCollins, 1997), 22.

17. Ibid., 15.

18. Carr, *Shallows*, 224.

19. Shenk, *Data Smog*, 52.

20. Charles Handy, "As I See It (the Future of Information)," in *i in the Sky: Visions of the Information Future*, ed. Alison Scammell (Chicago: Fitzroy Dearborn, 2000), 17.

21. Alan Schwartz, David M. Grether, and Louis L. Wilde, "The Irrelevance of Information Overload: An Analysis of Search and Disclosure," Yale Faculty Scholarship Series Paper 1123, 1986, pp. 277, 301, http://digitalcommons.law.yale.edu/fss_papers/1123.

22. Michael Webb, *Creating a New Reality: Information Age Effects on the Deception Process* (Maxwell Air Force Base, Alabama: Air University School of Advanced Air and Space Studies, 2006), 71.

23. "Fake News."

24. Richards J. Heuer, *Psychology of Intelligence Analysis* (Washington, DC: Central Intelligence Agency Center for the Study of Intelligence, 1999), 52.

25. Cynthia M. Grabo, *Anticipating Surprise: Analysis for Strategic Warning* (Bethesda, MD: Center for Strategic Intelligence Research at the Joint Military Intelligence College, 2002), 93.

26. Shenk, *Data Smog*, 15, 19.

27. "Too Much Information."

28. Grabo, *Anticipating Surprise*, 93.

29. Michael D. Brice, *Strategic Surprise in an Age of Information Superiority: Is It Still Possible?* (Maxwell Air Force Base, Alabama: Air War College at Air University, 2003), 4.

30. Michael I. Handel, "The Yom Kippur War and the Inevitability of Surprise," *International Studies Quarterly* 21 (September 1977): 464–468; Yigal Sheffy, "Overcoming

Strategic Weakness: The Egyptian Deception and the Yom Kippur War," *Intelligence and National Security* 21 (2006): 813–815.

31. Amir Oren, "The Yom Kippur War Came as a Surprise to the Egyptians as Well, Documents Show," *Haaretz*, October 13, 2016, https://www.haaretz.com/israel-news/ .premium-1973-war-came-as-surprise-to-egypt-as-well-documents-show-1.5448940.

32. Rodney C. Richardson, "Yom Kippur War: Grand Deception or Intelligence Blunder," *GlobalSecurity.org*, 1991, https://www.globalsecurity.org/military/library/ report/1991/RRC.htm.

33. Handel, "Yom Kippur War and the Inevitability of Surprise," 469.

34. Robert Mandel, *Optimizing Cyberdeterrence: A Comprehensive Strategy for Preventing Foreign Cyberattacks* (Washington, DC: Georgetown University Press, 2017), 88–94.

35. Handel, "Yom Kippur War and the Inevitability of Surprise," 473.

36. See Uri Bar-Joseph, *The Watchman Fell Asleep: The Surprise of the Yom Kippur War and Its Sources* (Albany: State University of New York Press, 2005).

37. Handel, "Yom Kippur War and the Inevitability of Surprise," 476.

38. Sheffy, "Overcoming Strategic Weakness," 825.

39. Handel, "Yom Kippur War and the Inevitability of Surprise," 462.

40. Michael J. Ahn, "Effective Policy Communication in the Age of Information Overload and YouTube," Brookings Institution, August 11, 2014, https://www .brookings.edu/blog/techtank/2014/08/11/effective-policy-communication-in-the-age -of-information-overload-and-youtube/.

41. Ibid.

42. "Fake News."

43. Will Oremus, "How 'Big Data' Went Bust," *Slate*, October 16, 2017, http://www .slate.com/articles/technology/technology/2017/10/what_happened_to_big_data.html.

44. Viktor Mayer-Schönberger and Kenneth Cukier, *Big Data: A Revolution That Will Transform How We Live, Work, and Think* (New York: Eamon Dolan/Mariner, 2014), 6.

45. Ibid., 12–13.

46. National Science Foundation, "Core Techniques and Technologies for Advancing Big Data Science and Engineering (BIGDATA)," 2012, http://www.nsf.gov/pubs/ 2012/nsf12499/nsf12499.pdf.

47. Kenneth Neil Cukier and Viktor Mayer-Schoenberger, "The Rise of Big Data: How It's Changing the Way We Think About the World," *Foreign Affairs* 92 (May–June 2013): 28.

48. Oremus, "How 'Big Data' Went Bust."

49. Cukier and Mayer-Schoenberger, "Rise of Big Data," 35.

50. Anthony G. Oettinger, "Whence and Whither Intelligence, Command and Control? The Certainty of Uncertainty," Harvard University Program on Information

Resources Policy, February 1990, p. 3, http://www.pirp.harvard.edu/pubs_pdf/oetting/
oetting-p90-1.pdf.

51. Kate Crawford, "The Hidden Biases of Big Data," *Harvard Business Review*, April
1, 2013, https://hbr.org/2013/04/the-hidden-biases-in-big-data.

52. Cukier and Mayer-Schoenberger, "Rise of Big Data," 31.

53. Mayer-Schönberger and Cukier, *Big Data*, 13, 32.

54. Megan Smith, D. J. Patil, and Cecilia Muñoz, "Big Risks, Big Opportunities:
The Intersection of Big Data and Civil Rights," May 4, 2016, https://obamawhitehouse
.archives.gov/blog/2016/05/04/big-risks-big-opportunities-intersection-big-data-and
-civil-rights.

55. Oremus, "How 'Big Data' Went Bust."

56. Damien Van Puyvelde, Shahriar Hossain, and Stephen Coulthart, "National Security Relies More and More on Big Data: Here's Why," *Washington Post*, September
27, 2017, https://www.washingtonpost.com/news/monkey-cage/wp/2017/09/27/national
-security-relies-more-and-more-on-big-data-heres-why.

57. Karan Jani, "The Promise and Prejudice of Big Data in Intelligence Community,"
October 26, 2016, p. 9, https://arxiv.org/ftp/arxiv/papers/1610/1610.08629.pdf.

58. Damien Van Puyvelde, Shahriar Hossain, and Stephen Coulthart, "Beyond the
Buzzword: Big Data and National Security Decision-Making," *International Affairs* 93
(2017): 19.

59. Jani, "The Promise and Prejudice of Big Data in Intelligence Community," 11–12.

60. Oremus, "How 'Big Data' Went Bust."

61. Erik Gartzke and Jon R. Lindsay, "Weaving Tangled Webs: Offense, Defense, and
Deception in Cyberspace," *Security Studies* 24 (2015): 333.

62. Kelvin Fu Wen Hao, "Globalisation and Its Impact on Military Intelligence,"
Pointer: Journal of the Singapore Armed Forces 41 (2015): 6.

63. Webb, *Creating a New Reality*, 71.

64. Jani, "The Promise and Prejudice of Big Data in Intelligence Community," 18.

65. Shenk, *Data Smog*, 29, 31.

66. "Too Much Information."

67. James Gleick, *The Information: A History, a Theory, a Flood* (New York: Pantheon, 2011), 403.

68. Gleick, *Information*, 403.

69. John B. Horrigan, "Information Overload," Pew Research Center, December 7,
2016, http://www.pewinternet.org/2016/12/07/information-overload/.

70. David Skyrme, "Information's Golden Age: Substance with Style," in *i in the
Sky: Visions of the Information Future*, ed. Alison Scammell (Chicago: Fitzroy Dearborn,
2000), 44.

71. Levitin, *Organized Mind*, 7.

72. Mayer-Schönberger and Cukier, *Big Data*, 9.

73. Nordenson, "Overload!"

74. Bill Kovach and Tom Rosenstiel, *Blur: How to Know What's True in the Age of Information Overload* (Bloomsbury, NJ: Bloomsbury USA, 2011), 6. See also Tyler Cowen, "How Real News Is Worse Than Fake News," *Bloomberg*, September 5, 2018, https://www.bloomberg.com/view/articles/2018-09-05/how-real-news-is-worse-than-fake-news.

75. Hao, "Globalisation and Its Impact on Military Intelligence," 3.

76. Ahn, "Effective Policy Communication."

77. Kovach and Rosenstiel, *Blur*, 7.

78. Gerald Sussman, *Communication, Technology, and Politics in the Information Age* (Thousand Oaks, CA: Sage, 1997), 51.

79. Illustrating such citizen fears are Cathy O'Neill, *Weapons of Math Destruction: How Big Data Increases Inequality and Threatens Democracy* (New York: Broadway, 2017); and Frank Pasquale, *The Black Box Society: The Secret Algorithms That Control Money and Information* (Cambridge, MA: Harvard University Press, 2016).

80. Erin M. Simpson, "The Peril and Promise of Big Data," *Foreign Policy*, May 15, 2014, http://foreignpolicy.com/2014/05/15/the-peril-and-promise-of-big-data/.

81. Simpson, "Peril and Promise of Big Data."

82. Mayer-Schönberger and Cukier, *Big Data*, 151.

83. Executive Office of the President, *Big Data: Seizing Opportunities, Preserving Values* (Washington, DC: U.S. Government Printing Office, 2014), 58–59.

84. Cukier and Mayer-Schoenberger, "Rise of Big Data," 37.

85. Vertzberger, *World in Their Minds*, 7; Michael I. Handel, *War, Strategy and Intelligence* (Totowa, NJ: Cass, 1989), 21.

86. Finel and Lord, "Surprising Logic of Transparency," 315–339.

87. Vertzberger, *World in Their Minds*, 1.

88. "Information Overload," *The Guardian*, January 16, 2000, https://www.theguardian.com/theobserver/2000/jan/16/newyou.life11.

89. Robert Jervis, *The Logic of Images in International Relations* (New York: Columbia University Press, 1989), 43, 46, 63.

90. Mayer-Schönberger and Cukier, *Big Data*, 73–97.

91. Shenk, *Data Smog*, 30.

92. Vertzberger, *World in Their Minds*, 31.

93. Kenneth J. Knapp, *Cyber Security and Global Information Assurance: Threat Analysis and Response Solutions* (Hershey, PA: Information Science Reference, 2009), 155.

94. Levitin, *Organized Mind*, 6.

95. Nassim Nicholas Taleb, *The Black Swan: The Impact of the Highly Improbable*, 2nd ed. (New York: Random House, 2010), xxii.

96. Steve Ragan, "Information Overload: Finding Signals in the Noise," *CSO*, May 29, 2014, http://www.csoonline.com/article/2243744/business-continuity/information-overload-finding-signals-in-the-noise.html.

97. Vertzberger, *World in Their Minds*, 28.

98. Ibid., 8.

99. Elias Groll, "Cyber Spying Is Out, Cyber Lying Is In," *Foreign Policy*, November 20, 2015, http://foreignpolicy.com/2015/11/20/u-s-fears-hackers-will-manipulate-data -not-just-steal-it/.

100. Ibid.

101. Karl von Clausewitz, *On War*, ed. Michael Howard and Peter Paret (Princeton, NJ: Princeton University Press, 1967), 198–199.

102. Abram Shulsky, "Elements of Strategic Denial and Deception," in *Strategic Denial and Deception: The Twenty-First Century Challenge*, ed. Roy Godson and James J. Wirtz (New Brunswick, NJ: Transaction, 2002), 19.

103. Abram Shulsky and Gary Schmitt, "Intelligence Reform: Beyond the Ames Case," in *U.S. Intelligence at the Crossroads: Agendas for Reform*, ed. Roy Godson, Ernest R. May, and Gary Schmitt (Washington, DC: Brassey's, 1995), 58.

104. Timothy J. Smith, "Overlord/Bodyguard: Intelligence Failure Through Adversary Deception," *International Journal of Intelligence and CounterIntelligence* 27 (2014): 550.

105. Sanne Blauw, "Why We Know Less About Developing Countries Than We Think," *The Correspondent*, June 23, 2016, https://thecorrespondent.com/4777/why-we -know-less-about-developing-countries-than-we-think/69518314778-cbc932ad.

106. Ibid.

107. Patricia Kingoni and René Gerrets, "Morals, Morale and Motivations in Data Fabrication: Medical Research Fieldworkers Views and Practices in Two Sub-Saharan African Contexts," *Social Science and Medicine* 166 (October 2016): 150–159.

108. Robert Mandel, *Dark Logic: Transnational Criminal Tactics and Global Security* (Stanford, CA: Stanford University Press, 2011), 91–126.

109. Blauw, "Why We Know Less About Developing Countries Than We Think."

110. Smith, "Overlord/Bodyguard," 551.

111. Ibid., 550, 564.

112. Grabo, *Anticipating Surprise*, 47, 164.

113. Robert Jervis, *Perception and Misperception in International Politics* (Princeton, NJ: Princeton University Press, 1976); Raymond S. Nickerson, "Confirmation Bias: A Ubiquitous Phenomenon in Many Guises," *Review of General Psychology* 2 (1998): 175–220.

114. See, for example, Robyn M. Dawes, "Shallow Psychology," in *Cognition and Social Behavior*, ed. J. Carroll and J. Payne (Hillsdale, NJ: Erlbaum, 1976), 7.

115. Ralph K. White, *Nobody Wanted War: Misperception in Vietnam and Other Wars* (Garden City, NY: Doubleday, 1970), chap. 8. For applications of black-and-white thinking, see T. McCartney, "American Nationalism and U.S. Foreign Policy from September

11 to the Iraq War," *Political Science Quarterly* 119 (Fall 2004): 408; Susan Page, "Poll of Several Countries Finds 'Complete Misperceptions,'" *USA Today*, June 23, 2006, p. 7; and Caspar W. Weinberger, "The Uses of Military Power," speech given to the National Press Club, Washington, DC, November 28, 1984, http://www.pbs.org/wgbh/pages/frontline/shows/military/force/weinberger.html.

116. See, for example, Jeff Victoroff, "The Mind of the Terrorist," *Journal of Conflict Resolution* 49 (February 2005): 56; Daniel Benjamin and Steven Simon, *The Next Attack: The Failure of the War on Terror and a Strategy for Getting It Right* (New York: Times Books, 2005), 74, 89; and Jerrold M. Post, "When Hatred Is Bred in the Bone: Psycho-Cultural Foundations of Contemporary Terrorism," *Political Psychology* 26 (August 2005): 628.

117. See Else Frenkel-Brunswik, "Intolerance of Ambiguity as an Emotional and Perceptual Personality Variable," *Journal of Personality* 18 (September 1949): 108–143; and Jamie Holmes, *Nonsense: The Power of Not Knowing* (New York: Crown, 2015), 11–12.

118. Stephen M. Walt, "Wishful Thinking: Top 10 Examples of the Most Unrealistic Expectations in Contemporary U.S. Foreign Policy," *Foreign Policy*, April 29, 2011, http://foreignpolicy.com/2011/04/29/wishful-thinking/.

119. Robert Mandel, "Adversaries Expectations and Desires About War Termination," in *Strategic War Termination*, ed. Stephen C. Cimbala (Westport, CT: Praeger, 1986), 177.

120. Robert Wuthnow, *Be Very Afraid: The Cultural Response to Terror, Pandemics, Environmental Devastation, Nuclear Annihilation, and Other Threats* (New York: Oxford University Press, 2010), 217.

121. Barry Buzan, *People, States and Fear: An Agenda for International Security Studies in the Post–Cold War Era*, 2nd ed. (Boulder, CO: Rienner, 1991), 140–141. See also Lewis M. Branscomb, "Vulnerability of Critical Infrastructure in the Twenty-First Century," in *Seeds of Disaster, Roots of Response: How Private Action Can Reduce Public Vulnerability*, ed. Philip E. Auerswald, Lewis M. Branscomb, Todd M. La Porte, and Erwann O. Michel-Kerjan (Cambridge: Cambridge University Press, 2006), 20.

122. Robert A. LeVine and Donald T. Campbell, *Ethnocentrism: Theories of Conflict, Ethnic Attitudes, and Group Behavior* (New York: Wiley, 1972), 212–223.

123. See Sam Keen, *Faces of the Enemy: Reflections of the Hostile Imagination* (New York: Harper and Row, 1986), 17; Jerome D. Frank and Andrei Y. Melville, "The Image of the Enemy and the Process of Change," in *Breakthrough—Emerging New Thinking: Soviet and Western Challenges Issue a Challenge to Build a World Beyond War*, ed. Anatoly Gromyko and Martin Hellman (New York: Walker, 1988), 199; and Katja M. Flückiger, "Xenophobia, Media Stereotyping, and Their Role in Global Insecurity," Geneva Centre for Security Policy Brief 21, December 6, 2006, https://www.files.ethz.ch/isn/92736/Brief-21.pdf.

124. Alison E. Smith, Lee Jussim, Jacquelynne Eccles, Michelle VanNoy, Stephanie Madon, and Polly Palumbo, "Self-Fulfilling Prophecies, Perceptual Biases, and Accuracy at the Individual and Group Levels," *Journal of Experimental Social Psychology* 34 (1998): 530–562.

125. Jervis, *Perception and Misperception in International Politics*, chap. 4.

126. Keren Yarhi-Milo, "Eye of the Beholder: How Leaders and Intelligence Communities Assess the Intentions of Adversaries," *International Security* 38 (Summer 2013): 9.

127. Arthur Gladstone, "The Conception of the Enemy," *Journal of Conflict Resolution* 3 (June 1959): 133.

128. LeVine and Campbell, *Ethnocentrism*, 212–223.

129. Philip E. Tetlock, "Social Psychology and World Politics," in *Handbook of Social Psychology*, 4th ed., ed. D. Gilbert, S. T. Fiske, and G. Lindsay (New York: McGraw-Hill, 2006), 868–914.

130. White, *Nobody Wanted War*, 303–305.

131. Oettinger, "Whence and Whither Intelligence, Command and Control?," 5.

132. Heuer, *Psychology of Intelligence Analysis*, xx.

133. Vertzberger, *World in Their Minds*, 111–113.

134. Jervis, *Perception and Misperception in International Politics*.

135. Irving L. Janis, *Groupthink* (New York: Free Press, 1982).

136. Janne E. Nolan and Douglas MacEachin, *Discourse, Dissent, and Strategic Surprise: Formulating U.S. Security Policy in an Age of Uncertainty* (Washington, DC: Georgetown University Institute for the Study of Diplomacy, 2006), 1, 2.

137. Bart Whaley, "Conditions Making for Success and Failure of Denial and Deception: Authoritarian and Transition Regimes," in *Strategic Denial and Deception: The Twenty-First Century Challenge*, ed. Roy Godson and James J. Wirtz (New Brunswick, NJ: Transaction, 2002), 78.

138. Klaus Knorr, "Threat Perception," in *Historical Dimensions of National Security Problems*, ed. Klaus Knorr (Lawrence: University Press of Kansas, 1976), 113.

139. See, for example, Morton Halperin, *Bureaucratic Politics and Foreign Policy* (Washington, DC: Brookings Institution, 1974).

140. Arie W. Kruglanski, Donna M. Webster, and Adena Klem, "Motivated Resistance and Openness to Persuasion in the Presence or Absence of Prior Information," *Journal of Personality and Social Psychology* 65 (1993): 861–876; Arie W. Kruglanski and Donna M. Webster, "Individual Differences in Need for Cognitive Closure," *Journal of Personality and Social Psychology* 67 (1994): 1049–1062.

141. Harold Wilensky, *Organizational Intelligence* (New York: Basic Books, 1967), 42–62, 126, 179; Richard K. Betts, "Analysis, War and Decision: Why Intelligence Failures Are Inevitable," *World Politics* 31 (October 1978): 67–68.

142. James J. Wirtz, "The Intelligence Paradigm," *Intelligence and National Security* 4 (October 1989): 829–837.

143. Herbert A. Simon, "Rational Choice and the Structure of the Environment," *Psychological Review* 63 (1966): 129–138.

144. Charles E. Parker and Eric K. Stern, "Bolt from the Blue or Avoidable Failure? Revisiting September 11 and the Origins of Strategic Surprise," *Foreign Policy Analysis* 1 (November 2005): 309.

145. Nolan and MacEachin, *Discourse, Dissent, and Strategic Surprise*, 2.

146. Grabo, *Anticipating Surprise*, 12.

147. Parker and Stern, "Bolt from the Blue or Avoidable Failure?," 310.

148. See Samuel Huntington, *The Clash of Civilizations and the Remaking of World Order* (New York: Simon and Schuster, 1996).

149. Magdalena Adriana Duvenage, "Intelligence Analysis in the Knowledge Age: An Analysis of the Challenges Facing the Practice of Intelligence Analysis" (master's thesis, Stellenbosch University, 2010), 50.

150. Oikonomakis Panagiotis, "Strategic Military Deception: Prerequisites of Success in Technological Environment," Research Institute for European and American Studies Research Paper No. 171, 2016, p. 26.

151. Heuer, *Psychology of Intelligence Analysis*, 70.

152. Grabo, *Anticipating Surprise*, iii.

153. Roy Godson and James J. Wirtz, "Strategic Denial and Deception," *International Journal of Intelligence and CounterIntelligence* 13 (2000): 426.

154. Heuer, *Psychology of Intelligence Analysis*, 34.

155. Ibid., 181.

156. This section draws heavily from Robert Mandel, *Deadly Transfers and the Global Playground: Transnational Security Threats in a Disorderly World* (Westport, CT: Praeger, 1999).

157. Max G. Manwaring and Courtney E. Prisk, "The Umbrella of Legitimacy," in *Gray Area Phenomena: Confronting the New World Disorder*, ed. Max G. Manwaring (Boulder, CO: Westview, 1993).

158. Zeev Maoz, *Paradoxes of War* (Boston: Unwin Hyman, 1990), 327.

159. Mikhail A. Alexseev, *Without Warning: Threat Assessment, Intelligence, and Global Struggle* (New York: St. Martin's, 1997), 262.

Chapter 2

1. Lani Kass and J. Phillip London, "Surprise, Deception, Denial and Warning: Strategic Imperatives," *Orbis* 57 (Winter 2013): 59.

2. Ibid., 65.

3. See Robert Jervis, *The Logic of Images in International Relations* (New York: Columbia University Press, 1989).

4. Abram Shulsky, "Elements of Strategic Denial and Deception," in *Strategic Denial and Deception: The Twenty-First Century Challenge*, ed. Roy Godson and James J. Wirtz

(New Brunswick, NJ: Transaction, 2002), 17. For more on the "deeply strategic character of public communication," see Piotr Cap, *The Language of Fear: Communicating Threat in Public Discourse* (London: Palgrave Macmillan, 2017), 81.

5. Jamie Holmes, *Nonsense: The Power of Not Knowing* (New York: Crown, 2015), 9.

6. Yaacov Y. I. Vertzberger, *The World in Their Minds: Information Processing, Cognition, and Perception in Foreign Policy Decisionmaking* (Stanford, CA: Stanford University Press, 1990), 32.

7. Donald C. Daniel and Katherine L. Herbig, "Propositions on Military Deception," in *Strategic Military Deception*, ed. Donald C. Daniel and Katherine L. Herbig (Elmsford, NY: Pergamon, 1981), 3.

8. Roy Godson and James J. Wirtz, "Strategic Denial and Deception," *International Journal of Intelligence and CounterIntelligence* 13 (2000): 424.

9. Kass and London, "Surprise, Deception, Denial and Warning," 61, 68.

10. Robert Mandel, *Optimizing Cyberdeterrence: A Comprehensive Strategy for Preventing Foreign Cyberattacks* (Washington, DC: Georgetown University Press, 2017), 196.

11. Kenneth Allen Cogswell, "Effects of Stress and Information Overload on Meaning Formation" (master's thesis, University of Montana, 1985), 2–3, http://scholarworks .umt.edu/cgi/viewcontent.cgi?article=5937&context=etd.

12. Magda Osman, "How Rational Is Deception?" July 2001, https://core.ac.uk/ download/pdf/30694870.pdf.

13. M. R. D. Foot, "Conditions Making for Success and Failure of Denial and Deception: Democratic Regimes," in *Strategic Denial and Deception: The Twenty-First Century Challenge*, ed. Roy Godson and James J. Wirtz (New Brunswick, NJ: Transaction, 2002), 122.

14. Kass and London, "Surprise, Deception, Denial and Warning," 75.

15. Michael Webb, *Creating a New Reality: Information Age Effects on the Deception Process* (Maxwell Air Force Base, Alabama: Air University School of Advanced Air and Space Studies, June 2006), 1.

16. Vertzberger, *World in Their Minds*, 31.

17. Richard K. Betts, "Analysis, War and Decision: Why Intelligence Failures Are Inevitable," *World Politics* 31 (October 1978): 69 (emphasis in original).

18. Ibid., 70.

19. Vertzberger, *World in Their Minds*, 215.

20. Cynthia M. Grabo, *Anticipating Surprise: Analysis for Strategic Warning* (Bethesda, MD: Center for Strategic Intelligence Research at the Joint Military Intelligence College, 2002), 46.

21. Betts, "Analysis, War and Decision," 70.

22. Grabo, *Anticipating Surprise*, 45.

23. Ellen Nakashima, "To Thwart Hackers, Firms Salting Their Servers with Fake Data," *Washington Post*, January 2, 2013, http://articles.washingtonpost.com/2013-01-02/ world/36211654_1_hackers-servers-contract-negotiations.

24. Amos Perlmutter and John Gooch, "Introduction," in *Military Deception and Strategic Surprise*, ed. John Gooch and Amos Perlmutter (Totawa, NJ: Cass, 1982), 1.

25. Webb, *Creating a New Reality*, 72.

26. Ibid., 50. See also Godson and Wirtz, "Strategic Denial and Deception," 431.

27. Eliot Cohen, "A Tale of Two Secretaries," *Foreign Affairs* 81 (May–June 2002): 42.

28. Kelvin Fu Wen Hao, "Globalisation and Its Impact on Military Intelligence," *Pointer: Journal of the Singapore Armed Forces* 41 (2015): 4.

29. Ephraim Kam, *Surprise Attack: The Victim's Perspective* (Cambridge, MA: Harvard University Press, 1988), 53.

30. Ibid., 54, 55.

31. Ibid., 37–38.

32. Janne E. Nolan and Douglas MacEachin, *Discourse, Dissent, and Strategic Surprise: Formulating U.S. Security Policy in an Age of Uncertainty* (Washington, DC: Georgetown University Institute for the Study of Diplomacy, 2006), 2.

33. Barry R. Schneider, "Deterring International Rivals from War and Escalation," in *Know Thy Enemy: Profiles of Adversary Leaders and Their Strategic Cultures*, ed. Barry R. Schneider and Jerrold M. Post (Maxwell Air Force Base, AL: United States Air Force Counterproliferation Center, 2003), 1.

34. Barton Whaley, "Stratagem: Deception and Surprise in War" (unpublished manuscript, Massachusetts Institute of Technology, Center for International Studies, 1969), 135.

35. Oikonomakis Panagiotis, "Strategic Military Deception: Prerequisites of Success in Technological Environment," Research Institute for European and American Studies Research Paper No. 171, 2016, p. 23.

36. Joint Chiefs of Staff, "Military Deception," Joint Publication 3-13.4, July 13, 2006, p. I-3.

37. For discussion of offensive vs. defensive preparations, see Grabo, *Anticipating Surprise*, 73–74.

38. Jervis, *Logic of Images in International Relations*, 123.

39. Florian P. Kühn, "Post-Interventionist *Zeitgeist*: The Ambiguity of Security Policy," in *The Armed Forces: Towards a Post-Interventionist Era?*, ed. Gerhard Kümmel and Bastian Giegerich (New York: Springer, 2013), 22.

40. Vertzberger, *World in Their Minds*, 343.

41. Perlmutter and Gooch, "Introduction," 1.

42. Richard K. Betts, *Surprise Attack: Lessons for Defense Planning* (Washington, DC: Brookings Institution, 1982), 109.

43. Michael I. Handel, "Intelligence and Deception," in *Military Deception and Strategic Surprise*, ed. John Gooch and Amos Perlmutter (Totawa, NJ: Cass, 1982), 124.

44. Jack Davis, "Strategic Warning: If Surprise Is Inevitable, What Role for Analysis?" *Kent Center Occasional Papers* 2, no. 1 (2003): 3.

45. Ibid.

46. Steve Chan, "The Intelligence of Stupidity: Understanding Failures in Strategic Warning," *American Political Science Review* 73 (March 1979): 173.

47. Betts, "Analysis, War and Decision," 70.

48. Jervis, *Logic of Images in International Relations*, 123.

49. Michael I. Handel, *Perception, Deception and Surprise: The Case of the Yom Kippur War* (Jerusalem: Leonard Davis Institute, 1976), 15.

50. Kass and London, "Surprise, Deception, Denial and Warning," 61, 68.

51. Betts, *Surprise Attack*, 2.

52. Ibid., 1.

53. Webb, *Creating a New Reality*, 71–72.

54. Robert Mandel, *Irrationality in International Confrontation* (Westport, CT: Greenwood, 1987), 4–6, 104, 105, 109.

55. Timothy J. Smith, "Overlord/Bodyguard: Intelligence Failure Through Adversary Deception," *International Journal of Intelligence and CounterIntelligence* 27 (2014): 564.

56. Richards J. Heuer, *Psychology of Intelligence Analysis* (Washington, DC: Central Intelligence Agency Center for the Study of Intelligence, 1999), 115.

57. Robert Jervis, *Perception and Misperception in International Politics* (Princeton, NJ: Princeton University Press, 1976), 135–137.

58. Ibid., 34–36.

59. Jervis, *Logic of Images in International Relations*, 115–116.

60. William Flavin, "Planning for Conflict Termination and Post-conflict Success," *Parameters* 32 (Autumn 2003): 97–98.

61. Shulsky, "Elements of Strategic Denial and Deception," 34.

62. Osman, "How Rational Is Deception?"

63. Handel, "Intelligence and Deception," 124–126.

64. Daniel and Herbig, "Propositions on Military Deception," 21–24.

65. Grabo, *Anticipating Surprise*, 122.

66. Godson and Wirtz, "Strategic Denial and Deception," 424.

67. Ibid., 425.

68. Grabo, *Anticipating Surprise*, 119.

69. Mandel, *Optimizing Cyberdeterrence*, 55–56.

70. Handel, "Intelligence and Deception," 139.

71. Thomas C. Schelling, "Foreword," in Rebecca Wohlstetter, *Pearl Harbor: Warning and Decision* (Stanford, CA: Stanford University Press, 1962), viii.

72. Grabo, *Anticipating Surprise*, 109.

73. Betts, *Surprise Attack*, 3.

74. Ibid., ix.

75. Robert Mandel, *Coercing Compliance: State-Initiated Brute Force in Today's World* (Stanford, CA: Stanford University Press, 2015), 5.

76. Stephen M. Walt, "Are Surprise Attacks a Thing of the Past?" *Foreign Policy*, March 8, 2012, http://foreignpolicy.com/2012/03/08/are-surprise-attacks-a-thing-of-the-past/.

77. Ibid.

78. Michael I. Handel, "Intelligence and the Problem of Strategic Surprise," *Journal of Strategic Studies* 7 (1984): 233.

79. Robert Mandel, *Deadly Transfers and the Global Playground: Transnational Security Threats in a Disorderly World* (Westport, CT: Praeger, 1999), 111–112.

80. Ibid., 112.

81. See, for example, Mandel, *Optimizing Cyberdeterrence*, 203.

82. Bart Whaley and Jeffrey Busby, "Detecting Deception: Practice, Practitioners, and Theory," in *Strategic Denial and Deception: The Twenty-First Century Challenge*, ed. Roy Godson and James J. Wirtz (New Brunswick, NJ: Transaction, 2002), 201.

83. Kenneth Geers, *Strategic Cyber Security* (Tallinn, Estonia: NATO Cooperative Cyber Defence Centre of Excellence, 2011), 100.

84. Irving Lachow, *Active Cyber Defense: A Framework for Policy Makers* (Washington, DC: Center for a New American Security, 2013), 6.

85. Martin Libicki, "Pulling Punches in Cyberspace," in *Proceedings of a Workshop on Deterring CyberAttacks: Informing Strategies and Developing Options for U.S. Policy*, ed. National Research Council (Washington, DC: National Academies Press, 2010), 125.

86. Martin Libicki, *Cyberdeterrence and Cyberwar* (Santa Monica, CA: Rand, 2009), 172.

87. Grabo, *Anticipating Surprise*, 129.

88. Ibid.

89. Alberto R. Coll, "Unconventional Warfare, Liberal Democracies, and International Order," *International Law Studies* 67 (1992): 17.

90. Godson and Wirtz, "Strategic Denial and Deception," 429.

91. Foot, "Conditions Making for Success and Failure of Denial and Deception," 116–117.

92. Godson and Wirtz, "Strategic Denial and Deception," 427.

93. J. Bowyer Bell, "Conditions Making for Success and Failure of Denial and Deception: Nonstate and Illicit Actors," in *Strategic Denial and Deception: The Twenty-First Century Challenge*, ed. Roy Godson and James J. Wirtz (New Brunswick, NJ: Transaction, 2002), 129.

94. Hao, "Globalisation and Its Impact on Military Intelligence," 6.

95. Nolan and MacEachin, *Discourse, Dissent, and Strategic Surprise*, 1.

96. Betts, *Surprise Attack*, 16–17.

97. Robert Mandel, "On Estimating Post–Cold War Enemy Intentions," *Intelligence and National Security* 24 (April 2009): 194–215.

98. Nolan and MacEachin, *Discourse, Dissent, and Strategic Surprise*, 2.

99. Vertzberger, *World in Their Minds*, 385.

100. Mandel, *Irrationality in International Confrontation*, 104, 105, 111.

101. Jervis, *Logic of Images in International Relations*, 130–131.

102. Arnold Wolfers, "'National Security' as an Ambiguous Symbol," *Political Science Quarterly* 67 (December 1952): 481. See also Robert Mandel, "What Are We Protecting?" *Armed Forces and Society* 22 (Spring 1996): 335–355.

103. Jervis, *Perception and Misperception in International Politics*, chap. 4.

104. Chan, "Intelligence of Stupidity," 172.

105. Mandel, *Deadly Transfers and the Global Playground*, 109.

106. Robert Mandel, *Global Security Upheaval: Armed Nonstate Groups Usurping State Stability Functions* (Stanford, CA: Stanford University Press, 2013), 189.

107. Lawrence Freedman, "Order and Disorder in the New World," *Foreign Affairs* 71 (1991–1992): 30.

108. Hong Kian Wah, "Irregular Warfare," *Pointer: Journal of the Singapore Armed Forces* 26 (2000), previously available at http://www.mindef.gov.sg/safti/pointer/back/journals/2000/Vol26_3/7.htm.

109. Robert Mandel, *Dark Logic: Transnational Criminal Tactics and Global Security* (Stanford, CA: Stanford University Press, 2011), 157.

110. Erik Gartzke and Jon R. Lindsay, "Weaving Tangled Webs: Offense, Defense, and Deception in Cyberspace," *Security Studies* 24 (2015): 338–339; Nakashima, "To Thwart Hackers."

111. Erping Zhang, "SARS: Unmasking Censorship in China," Association for Asian Research, August 11, 2003, http://www.asianresearch.org/articles/1502.html.

112. Andrew J. Bacevich, Max Boot, Michael Ignatieff, Michael O'Hanlon, and Jonathan Masters, "Was the Iraq War Worth It?" Council on Foreign Relations, December 15, 2011, http://www.cfr.org/iraq/iraq-war-worth-/p26820.

113. Vertzberger, *World in Their Minds*, 29.

114. Mandel, *Optimizing Cyberdeterrence*, 241.

115. James J. Wirtz, "Theory of Surprise," in *Paradoxes of Strategic Intelligence: Essays in Honor of Michael I. Handel*, ed. Richard K. Betts and Thomas G. Mahnken (Portland, OR: Cass, 2003), 109–110.

116. Jervis, *Logic of Images in International Relations*, 123.

117. Grabo, *Anticipating Surprise*, 129.

118. Steven Metz and Raymond A. Millen, "Future War/Future Battlespace: The Strategic Role of American Landpower" (monograph, U.S. Army War College Strategic Studies Institute, 2003), 18.

119. Geoffrey Barker, "Strategic Ambiguity," *The Strategist*, October 20, 2015, https://www.aspistrategist.org.au/strategic-ambiguity/.

120. Robert Mandel, *Global Threat: Target-Centered Assessment and Management* (Westport, CT: Praeger Security International, 2008), 64.

121. Joseph S. Nye Jr., *Cyber Power* (Cambridge, MA: Belfer Center for Science and International Affairs, 2010), 5; Benjamin Brake, "Strategic Risks of Ambiguity in Cyberspace," Council on Foreign Relations Contingency Planning Memorandum No. 24, May 14, 2015, https://www.cfr.org/report/strategic-risks-ambiguity-cyberspace.

122. For a discussion of the tensions surrounding false transparency, see Harry Farrell and Martha Finnemore, "The End of Hypocrisy: American Foreign Policy in the Age of Leaks," *Foreign Affairs* 92 (November–December 2013): 21.

123. Foot, "Conditions Making for Success and Failure of Denial and Deception," 127.

124. Whaley and Busby, "Detecting Deception," 185.

125. Grabo, *Anticipating Surprise*, 121.

126. Handel, "Intelligence and Deception," 122.

127. Ibid.

128. Handel, "Intelligence and the Problem of Strategic Surprise," 229–230.

129. Richard N. Haass, *Intervention: The Use of American Military Force in the Post–Cold War World* (Washington, DC: Carnegie Endowment for International Peace, 1994), 89.

Chapter 3

1. Quinta Jurecic, "Trump Administration Releases 'Briefing Issues' on Foreign Policy and Security," *Lawfare*, January 20, 2017, https://www.lawfareblog.com/trump-administration-releases-briefing-issues-foreign-policy-and-security.

2. William Roberts, "Decoding Donald Trump's Foreign Policy," *Al Jazeera*, June 5, 2017, http://www.aljazeera.com/news/2017/06/decoding-donald-trumps-foreign-policy-170605220035165.html.

3. Thomas Wright, "Trump's 19th Century Foreign Policy," *Politico Magazine*, January 20, 2016, http://www.politico.com/magazine/story/2016/01/donald-trump-foreign-policy-213546.

4. Jurecic, "Trump Administration Releases 'Briefing Issues.'"

5. Ibid.

6. Stephen Sestanovich, "The Brilliant Incoherence of Trump's Foreign Policy," *The Atlantic*, May 2017, https://www.theatlantic.com/magazine/archive/2017/05/the-brilliant-incoherence-of-trumps-foreign-policy/521430/.

7. Dani Nedal and Daniel Nexon, "Trump's 'Madman Theory' Isn't Strategic Unpredictability: It's Just Crazy," *Foreign Policy*, April 18, 2017, http://foreignpolicy.com/2017/04/18/trumps-madman-theory-isnt-strategic-unpredictability-its-just-crazy/.

8. Timothy P. Carney, "Strategic Ambiguity and Peace Through Strength: A Hopeful Interpretation of Trump on Syria," *Washington Examiner*, April 11, 2017, http://www.washingtonexaminer.com/strategic-ambiguity-and-peace-through-strength-a-hopeful-interpretation-of-trump-on-syria/article/2620035.

9. Kevin Sullivan and Karen Tumulty, "Trump Promised an 'Unpredictable' Foreign Policy: To Allies, It Looks Incoherent," *Washington Post*, April 11, 2017, https://www.washingtonpost.com/politics/trump-promised-an-unpredictable-foreign-policy-to-allies-it-looks-incoherent/2017/04/11/21acde5e-1a3d-11e7-9887-1a5314b56a08_story.html.

10. Glenn Thrush and Mark Landler, "Bold, Unpredictable Foreign Policy Lifts Trump, but Has Risks," *New York Times*, April 20, 2017, https://www.nytimes.com/2017/04/20/us/politics/trump-foreign-policy.html.

11. "Fake News," *60 Minutes*, March 26, 2017, https://www.cbsnews.com/video/the-attack-in-garland-fake-news-chess-country.

12. Wright, "Trump's 19th Century Foreign Policy."

13. Van Jackson, "Reading Trump: The Danger of Overanalyzing His Tweets," *Foreign Affairs*, January 25, 2017, https://www.foreignaffairs.com/articles/2017-01-25/reading-trump.

14. Ibid.

15. Wright, "Trump's 19th Century Foreign Policy."

16. Catherine Rampell, "Trump's Unpredictability May Be Useful Against Enemies, but It's Deadly to Allies," *Washington Post*, July 25, 2016, https://www.washingtonpost.com/opinions/trumps-unpredictability-may-be-useful-against-enemies-but-its-deadly-to-allies/2016/07/25/37019d86-52a0-11e6-88eb-7dda4e2f2aec_story.html.

17. Julie Pace, "Trump's Ambiguity on Foreign Policy a High-Risk Doctrine for President," Talking Points Memo, December 23, 2016, http://talkingpointsmemo.com/news/trump-ambiguity-foreign-policy-high-risk-doctrine.

18. Ibid.

19. Andy Keiser, "100 Days in, Trump's Foreign Policy Plays the Strategic Ambiguity Game," *The Hill*, April 28, 2017, http://thehill.com/blogs/pundits-blog/foreign-policy/331019-100-days-in-trumps-foreign-policy-plays-the-strategic.

20. David Ignatius, "Trump Uses Ambiguity in Foreign Policy," *Newsmax*, December 21, 2016, http://www.newsmax.com/DavidIgnatius/trump-china-russia-policy/2016/12/21/id/765011/.

21. Nedal and Nexon, "Trump's 'Madman Theory' Isn't Strategic Unpredictability."

22. Anne R. Pierce, "A Promising Start: President Trump's Foreign Policy Team Deserves Credit for Its Work on China and North Korea," *US News and World Report*, April 28, 2017, https://www.usnews.com/opinion/articles/2017-04-28/donald-trumps-foreign-policy-is-off-to-a-good-start.

23. Charles Krauthammer, "Trump's Great Reversal—for Now," *National Review*, April 14, 2017, http://www.nationalreview.com/article/446732/trump-foreign-policy-unpredictable-idiosyncratic.

24. Guy Taylor and Dan Boylan, "Trump's Unconventional Foreign Policy Makes Consequential Impact in First 100 Days," *Washington Times*, April 27, 2017, http://www.washingtontimes.com/news/2017/apr/27/donald-trumps-foreign-policy-makes-impact/.

25. Richard N. Haass, "Where to Go from Here: Rebooting American Foreign Policy," *Foreign Affairs* 96 (July–August 2017): 8.

26. Ignatius, "Trump Uses Ambiguity in Foreign Policy."

27. Rampell, "Trump's Unpredictability May be Useful Against Enemies."

28. Nedal and Nexon, "Trump's 'Madman Theory' Isn't Strategic Unpredictability."

29. Michael Young, "Obama Was Bad for the Middle East, but Trump's Ambiguity Is Worrisome," *The National*, July 26, 2017, https://www.thenational.ae/opinion/comment/obama-was-bad-for-the-middle-east-but-trump-s-ambiguity-is-worrisome-1.614416.

30. Ross Douthat, "The Fog of Trump," *New York Times*, January 28, 2017, https://www.nytimes.com/2017/01/28/opinion/sunday/the-fog-of-trump.html.

31. Nedal and Nexon, "Trump's 'Madman Theory' Isn't Strategic Unpredictability."

32. Joe Schneider, "How Trump's White House Is Upping Its War on Leaks," *Bloomberg Businessweek*, July 27, 2017, https://www.bloomberg.com/news/articles/2017-07-27/how-trump-white-house-is-upping-its-war-on-leaks-quicktake-q-a.

33. Michael H. Fuchs, "Donald Trump's Doctrine of Unpredictability Has the World on Edge," *The Guardian*, February 13, 2017, https://www.theguardian.com/commentisfree/2017/feb/13/donald-trumps-doctrine-unpredictability-world-edge.

34. Thrush and Landler, "Bold, Unpredictable Foreign Policy Lifts Trump."

35. Nedal and Nexon, "Trump's 'Madman Theory' Isn't Strategic Unpredictability."

36. Ibid.

37. Sullivan and Tumulty, "Trump Promised an 'Unpredictable' Foreign Policy."

38. Ibid.

39. Ibid.

40. Ibid.

41. Wright, "Trump's 19th Century Foreign Policy."

42. Jackson, "Reading Trump."

43. Andre van Loon, "From a Bang to a Whisper: The Decline in Brexit-Related Discussion on Social Media," London School of Economics and Political Science, October 31, 2016, http://blogs.lse.ac.uk/brexit/2016/10/31/from-a-bang-to-a-whisper-the-decline-in-brexit-related-discussion-on-social-media.

44. Oliver Daddow, "Delusions and Meddling: 30 Years of Tory Euroscepticism Are Coming to the Fore," London School of Economics and Political Science, May 9,

2017, http://blogs.lse.ac.uk/brexit/2017/05/09/delusions-and-meddling-30-years-of-tory-euroscepticism-are-coming-to-the-fore/.

45. Craig Oliver, *Unleashing Demons: The Inside Story of Brexit* (London: Hodder and Stoughton, 2016), 9.

46. Timothy Heppell, Andrew Crines, and David Jeffery, "The United Kingdom Referendum on European Union Membership: The Voting of Conservative Parliamentarians," *Journal of Common Market Studies*, 55 (2017): 768.

47. Ashley Kirk, "EU Referendum: The Claims That Won It for Brexit, Fact Checked," *The Telegraph*, March 13, 2017, http://www.telegraph.co.uk/news/0/eu-referendum-claims-won-brexit-fact-checked/.

48. Will Brett, "The People Have Spoken—or Have They? Doing Referendums Differently After the EU Vote," London School of Economics and Political Science, September 1, 2016, http://blogs.lse.ac.uk/brexit/2016/09/01/the-people-have-spoken-or-have-they-doing-referendums-differently-after-the-eu-vote/.

49. British Polling Council, "Performance of the Polls in the EU Referendum," June 24, 2016, http://www.britishpollingcouncil.org/performance-of-the-polls-in-the-eu-referendum/

50. Brett, "People Have Spoken."

51. Oliver, *Unleashing Demons*, 10.

52. Martin Moore and Gordon Ramsay, "Acrimonious and Divisive: The Role the Media Played in Brexit," London School of Economics and Political Science, May 16, 2017, http://blogs.lse.ac.uk/brexit/2017/05/16/acrimonious-and-divisive-the-role-the-media-played-in-brexit/.

53. Ibid.

54. Daddow, "Delusions and Meddling."

55. Van Loon, "From a Bang to a Whisper."

56. Moore and Ramsay, "Acrimonious and Divisive."

57. Van Loon, "From a Bang to a Whisper."

58. Moore and Ramsay, "Acrimonious and Divisive."

59. Ibid.

60. Piotr Cap, *The Language of Fear: Communicating Threat in Public Discourse* (London: Palgrave Macmillan, 2017), 67.

61. Moore and Ramsay. "Acrimonious and Divisive."

62. Kirk, "EU Referendum."

63. Peter Marshall, "Britain Is Fit for Triggering," *Round Table* 106 (2017): 212.

64. Heppell, Crines, and Jeffery, "United Kingdom Referendum on European Union Membership," 767.

65. Oliver, *Unleashing Demons*, 403.

66. See Andrew Glencross, *Why the UK Voted for Brexit: David Cameron's Great Miscalculation* (London: Palgrave Macmillan, 2016).

67. Heppell, Crines, and Jeffery, "United Kingdom Referendum on European Union Membership," 776.

68. Oliver, *Unleashing Demons*, 11.

69. Jakub Grygiel, "The Return of Europe's Nation-States: The Upside to the EU's Crisis," *Foreign Affairs* 95 (2016): 97.

70. Ibid., 98.

71. Harold James, "Brexit Fudge," Project Syndicate, August 2, 2016, https://www.project-syndicate.org/commentary/uk-post-brexit-relationship-eu-by-harold-james-2016-08.

72. Ibid.

73. Ibid.

74. Marshall, "Britain Is Fit for Triggering," 211.

75. Grygiel, "Return of Europe's Nation-States," 94.

76. Ibid., 95.

77. Richard Ekins, "The Legitimacy of the Brexit Referendum," UK Constitutional Law Association, June 29, 2016, https://ukconstitutionallaw.org/2016/06/29/richard-ekins-the-legitimacy-of-the-brexit-referendum/.

78. Ibid.

79. Davina Cooper, "Thinking Harder: How We Could Do Referendums Differently," London School of Economics and Political Science, November 18, 2016, http://blogs.lse.ac.uk/brexit/2016/11/18/thinking-harder-how-we-could-do-referendums-differently/.

80. "Legitimacy of the Brexit Referendum," *The Guardian*, July 12, 2016, https://www.theguardian.com/politics/2016/jul/12/legitimacy-of-the-brexit-referendum.

81. Moore and Ramsay. "Acrimonious and Divisive."

82. Ibid.

83. Daddow, "Delusions and Meddling."

84. Moore and Ramsay, "Acrimonious and Divisive."

85. Daddow, "Delusions and Meddling."

86. Grygiel, "Return of Europe's Nation-States," 94.

87. James, "Brexit Fudge."

88. Oliver, *Unleashing Demons*, 404.

89. Liesbeth Van der Heide, "Cherry-Picked Intelligence: The Weapons of Mass Destruction Dispositive as a Legitimation for National Security in the Post 9/11 Age," *Historical Social Research* 38 (2013): 286.

90. Ibid., 287.

91. Ibid.

92. Marcus Warren, "Don't Mess with Us, UN Warns Saddam," *The Telegraph*, November 16, 2002, http://www.telegraph.co.uk/news/worldnews/northamerica/usa/1413384/Dont-mess-with-us-UN-warns-Saddam.html.

93. Will Knight, "UN Weapons Inspectors Told to Leave Iraq," *New Scientist*, March 17, 2003, http://www.newscientist.com/article/dn3509-un-weapons-inspectors told-to -leave-iraq.html.

94. Van der Heide, "Cherry-Picked Intelligence," 286.

95. Kurt Eichenwald, "The War on Error," *Newsweek Global*, July 29, 2016, p. 12.

96. Wright Bryan, "Iraq WMD Timeline: How the Mystery Unraveled," *NPR*, November 15, 2005, http://www.npr.org/templates/story/story.php?storyId=4996218.

97. Kenneth M. Pollack, "Spies, Lies, and Weapons: What Went Wrong," *The Atlantic*, January–February 2004, https://www.theatlantic.com/magazine/archive/2004/01/ spies-lies-and-weapons-what-went-wrong/302878.

98. Van der Heide, "Cherry-Picked Intelligence," 290.

99. Ibid.

100. Charles A. Duelfer and Stephen Benedict Dyson, "Chronic Misperception and International Conflict: The U.S.-Iraq Experience," *International Security* 36 (Summer 2011): 97.

101. Mark Strauss, "Attacking Iraq," *Foreign Policy*, no. 129 (March–April 2002): 16.

102. Duelfer and Dyson, "Chronic Misperception and International Conflict," 98.

103. Evan Thomas and Karen Breslau, "Saddam's Dark Threat," *Newsweek*, November 24, 1997, p. 24.

104. Van der Heide, "Cherry-Picked Intelligence," 290.

105. John Schwarz, "Twelve Years Later, US Media Still Can't Get Iraqi WMD Story Right," *The Intercept*, April 10, 2015, https://theintercept.com/2015/04/10/twelve-years -later-u-s-media-still-cant-get-iraqi-wmd-story-right/.

106. "'No Surprise' Iraq WMD Not Found," *BBC News*, October 3, 2003, http://news .bbc.co.uk/2/hi/americas/3160602.stm.

107. Ibid.

108. Bryan, "Iraq WMD Timeline: How the Mystery Unraveled."

109. "'No Surprise' Iraq WMD Not Found."

110. Van der Heide, "Cherry-Picked Intelligence," 293.

111. Pollack, "Spies, Lies, and Weapons."

112. Van der Heide, "Cherry-Picked Intelligence," 288.

113. Duelfer and Dyson, "Chronic Misperception and International Conflict," 78.

114. Ibid., 99.

115. Michael Eisenstadt, "Understanding Saddam," *National Interest*, no. 81 (2005): 119.

116. Duelfer and Dyson, "Chronic Misperception and International Conflict," 73.

117. Eisenstadt, "Understanding Saddam," 118.

118. Ibid.

119. Duelfer and Dyson, "Chronic Misperception and International Conflict," 74.

120. Van der Heide, "Cherry-Picked Intelligence," 289, 299.

121. Bruce Riedel, "Lessons of the Syrian Reactor," *National Interest*, no. 125 (May–June 2013): 45.

122. Duelfer and Dyson, "Chronic Misperception and International Conflict," 97.

123. Van der Heide, "Cherry-Picked Intelligence," 299.

124. Eichenwald, "War on Error," 15.

125. Michael Duffy, "Weapons of Mass Disappearance," *Time*, June 9, 2003, p. 28.

126. "O'Neill: 'Frenzy' Distorted War Plans Account," *CNN*, January 14, 2004, http://www.cnn.com/2004/ALLPOLITICS/01/13/oneill.bush.

127. Pollack, "Spies, Lies, and Weapons."

128. Ibid.

129. Ibid. (emphasis in original).

130. Eisenstadt, "Understanding Saddam," 119.

131. Van der Heide, "Cherry-Picked Intelligence," 302.

132. Walter Pincus, "Ex-CIA Official Faults Use of Data on Iraq," *Washington Post*, February 10, 2006, http://www.washingtonpost.com/wp-dyn/content/article/2006/02/09/AR2006020902418.html.

133. Duelfer and Dyson, "Chronic Misperception and International Conflict," 74.

134. Ibid., 96.

135. Bob Drogin, "The Vanishing," *New Republic*, July 21, 2003, p. 21.

136. Ibid.

137. Eisenstadt, "Understanding Saddam," 119.

138. Pollack, "Spies, Lies, and Weapons."

139. Van der Heide, "Cherry-Picked Intelligence," 297–298.

140. Glenn Kessler, "The Pre-war Intelligence on Iraq: Wrong or Hyped by the Bush White House?" *Washington Post*, December 13, 2016, https://www.washingtonpost.com/news/fact-checker/wp/2016/12/13/the-pre-war-intelligence-on-iraq-wrong-or-hyped-by-the-bush-white-house.

141. Kathy Gannon, "Afghanistan Unbound," *Foreign Affairs* 83 (May–June 2004): 41.

142. Van der Heide, "Cherry-Picked Intelligence," 298.

143. Eichenwald, "War on Error," 17.

144. Riedel, "Lessons of the Syrian Reactor," 45.

145. "'No Surprise' Iraq WMD Not Found."

146. Van der Heide, "Cherry-Picked Intelligence," 297.

147. Pollack, "Spies, Lies, and Weapons."

148. Duffy, "Weapons of Mass Disappearance." 28.

149. Kessler, "Prewar Intelligence on Iraq."

150. Eichenwald, "War on Error," 16.

151. John Barry and Mark Hosenball, "What Went Wrong," *Newsweek*, February 9, 2004, p. 25.

152. Roy Allison, "Russian 'Deniable' Intervention in Ukraine: How and Why Russia Broke the Rules," *International Affairs* 90 (2014): 1255.

153. Anton Bebler, "Crimea and the Russian-Ukrainian Conflict," *Romanian Journal of European Affairs* 15 (2015): 37.

154. Ibid., 41–42.

155. Johan Norberg, "The Use of Russia's Military in the Crimean Crisis," Carnegie Endowment for International Peace, March 13, 2014, http://carnegieendowment.org/2014/03/13/use-of-russia-s-military-in-crimean-crisis-pub-54949.

156. "Ukraine's Petro Poroshenko Pledges 'End to War,'" *BBC News*, May 26, 2014, http://www.bbc.com/news/world-europe-27571612.

157. Sascha-Dominik Bachmann and Håkan Gunneriusson, "Hybrid Wars: 21st Century's New Threats to Global Peace and Security," *South African Journal of Military Studies* 43 (2015): 88.

158. Jeffrey Mankoff, "Russia's Latest Land Grab: How Putin Won Crimea and Lost Ukraine," *Foreign Affairs* 93 (May–June 2014): 60.

159. Nikolas K. Gvosdev, "Ukraine and the Failure of Strategic Ambiguity," *National Interest*, March 4, 2014, http://nationalinterest.org/commentary/ukraine-the-failure-strategic-ambiguity-9992.

160. Bebler, "Crimea and the Russian-Ukrainian Conflict," 49.

161. Ibid., 51–52.

162. Ibid., 41.

163. Kristin Ven Bruusgaard, "Crimea and Russia's Strategic Overhaul," *Parameters* 44 (2014): 83.

164. Bebler, "Crimea and the Russian-Ukrainian Conflict," 41.

165. Allison, "Russian 'Deniable' Intervention in Ukraine," 1258.

166. Peter Pomerantsev, "Inside the Kremlin's Hall of Mirrors," *The Guardian*, April 9, 2015, https://www.theguardian.com/news/2015/apr/09/kremlin-hall-of-mirrors-military-information-psychology.

167. Arkady Ostrovsky, "For Putin, Disinformation Is Power," *New York Times*, August 5, 2016, https://www.nytimes.com/2016/08/06/opinion/for-putin-disinformation-is-power.html.

168. Allison, "Russian 'Deniable' Intervention in Ukraine," 1259.

169. "A Guide to Russian Propaganda, Part 1: Propaganda Prepares Russia for War," Euromaidan Press, May 5, 2016, http://euromaidanpress.com/2016/05/05/a-guide-to-russian-propaganda-part-1-propaganda-prepares-russia-for-war/#arvlbdata.

170. Lucy Ash, "How Russia Outfoxes Its Enemies," *BBC News*, January 29, 2015, http://www.bbc.com/news/magazine-31020283.

171. Ibid.

172. Daniel Treisman, "Why Putin Took Crimea: The Gambler in the Kremlin," *Foreign Affairs* (May–June 2016): 54.

173. Magnus Christiansson, *Strategic Surprise in the Ukraine Crisis* (Stockholm, Sweden: Swedish National Defence College, 2014), 5, 61.

174. Bebler, "Crimea and the Russian-Ukrainian Conflict," 49.

175. Christiansson, *Strategic Surprise in the Ukraine Crisis*, 4.

176. Bebler, "Crimea and the Russian-Ukrainian Conflict," 49.

177. Reuters, "NATO: Russian Propaganda Up Since Crimea Offensive," *Newsweek*, February 11, 2017, http://www.newsweek.com/nato-russian-propaganda-crimea-offensive-555724.

178. Pomerantsev, "Inside the Kremlin's Hall of Mirrors."

179. Peter Pomerantsev, "The Kremlin's Information War," *Journal of Democracy* 26 (October 2015): 43.

180. Pomerantsev, "Inside the Kremlin's Hall of Mirrors."

181. Pomerantsev, "Kremlin's Information War," 42.

182. Ostrovsky, "For Putin, Disinformation Is Power."

183. Thomas Mehlhausen, "Which Carrots and Which Sticks? Searching for a Strategy to Deal with Russia in the Ukraine Crisis," American Institute for Contemporary German Studies, May 11, 2015, http://www.aicgs.org/publication/which-carrots-and-which-sticks/.

184. Pomerantsev, "Inside the Kremlin's Hall of Mirrors."

185. Jolanta Darczewska, "The Anatomy of Russian Information Warfare: The Crimean Operation, A Case Study," *Point of View* 42 (May 2014):34, https://www.osw.waw.pl/sites/default/files/the_anatomy_of_russian_information_warfare.pdf.

186. Ostrovsky, "For Putin, Disinformation Is Power."

187. Allison, "Russian 'Deniable' Intervention in Ukraine," 1268–1269, 1295–1296.

188. Sophie Pinkham, "How Annexing Crimea Allowed Putin to Claim He Had Made Russia Great Again," *The Guardian*, March 22, 2017, https://www.theguardian.com/commentisfree/2017/mar/22/annexing-crimea-putin-make-russia-great-again.

189. Ibid.

190. Bebler, "Crimea and the Russian-Ukrainian Conflict," 39–40.

191. Ibid., 40.

192. Allison, "Russian 'Deniable' Intervention in Ukraine," 1296.

193. Bebler, "Crimea and the Russian-Ukrainian Conflict," 41.

194. Mankoff, "Russia's Latest Land Grab," 67.

195. Treisman, "Why Putin Took Crimea," 48.

196. Gvosdev, "Ukraine and the Failure of Strategic Ambiguity."

197. Bebler, "Crimea and the Russian-Ukrainian Conflict," 48.

198. Ven Bruusgaard, "Crimea and Russia's Strategic Overhaul," 86.

199. Pinkham, "How Annexing Crimea Allowed Putin to Claim He Had Made Russia Great Again."

200. Treisman, "Why Putin Took Crimea," 52.

201. Mankoff, "Russia's Latest Land Grab," 61.

202. Gvosdev, "Ukraine and the Failure of Strategic Ambiguity."

203. Ibid.

204. Ivan Krastev and Mark Leonard, "Europe's Shattered Dream of Order: How Putin Is Disrupting the Atlantic Alliance," *Foreign Affairs*, May–June 2015, https://www.foreignaffairs.com/articles/western-europe/2015-04-20/europes-shattered-dream-order.

205. Bebler, "Crimea and the Russian-Ukrainian Conflict," 50.

206. Allison, "Russian 'Deniable' Intervention in Ukraine," 1255.

207. Pinkham, "How Annexing Crimea Allowed Putin to Claim He Had Made Russia Great Again."

208. Krastev and Leonard, "Europe's Shattered Dream of Order."

209. Bebler, "Crimea and the Russian-Ukrainian Conflict," 45.

210. Allison, "Russian 'Deniable' Intervention in Ukraine," 1258.

211. Pinkham, "How Annexing Crimea Allowed Putin to Claim He Had Made Russia Great Again."

212. Pomerantsev, "Kremlin's Information War," 48.

213. Bebler, "Crimea and the Russian-Ukrainian Conflict," 46–47.

214. Bebler, "Crimea and the Russian-Ukrainian Conflict," 51.

215. Bernhard Stahl, Robin Lucke, and Anna Felfeli, "Comeback of the Transatlantic Security Community? Comparative Securitisation in the Crimea Crisis," *East European Politics* 32 (2016): 525–526, 535.

216. Ibid., 538.

217. Krastev and Leonard, "Europe's Shattered Dream of Order."

218. Dmitri Trenin, "The Revival of the Russian Military: How Moscow Reloaded," *Foreign Affairs* 95 (May–June 2016): 29; Treisman, "Why Putin Took Crimea," 49.

219. Bebler, "Crimea and the Russian-Ukrainian Conflict," 51.

220. Allison, "Russian 'Deniable' Intervention in Ukraine," 1297.

221. Richard J. Samuels, *3.11: Disaster and Change in Japan* (Ithaca, NY: Cornell University Press, 2013), ix.

222. Itoko Suzuki and Yuko Kaneko, *Japan's Disaster Governance: How Was the 3.11 Crisis Managed?* (New York: Springer, 2012), 7.

223. Tom Gill, Brigitte Steger, and David H. Slater, "The 3.11 Disasters," in *Japan Copes with Calamity: Ethnographies of the Earthquake, Tsunami and Nuclear Disasters of March 2011*, ed. Tom Gill, Brigitte Steger, and David H. Slater (Oxford, UK: Lang, 2013), 5.

224. Daniel Kaufmann and Veronika Penciakova, "Japan's Triple Disaster: Governance and the Earthquake, Tsunami and Nuclear Crises," Brookings, March 16, 2011, http://www.brookings.edu/research/opinions/2011/03/16-japan-disaster-kaufmann.

225. Gill, Steger, and Slater, "3.11 Disasters," 5.

226. "Fukushima 2013: A Continuing Nuclear Disaster of Global Significance," *Truth11*, March 23, 2013, http://truth11.com/2013/03/23/fukushima2013acontinuing nucleardisasterofglobalsignificance/.

227. Kaufmann and Penciakova, "Japan's Triple Disaster."

228. Initiative for Global Environmental Leadership, "Disasters, Leadership, and Rebuilding—Tough Lessons from Japan and the U.S.," October 2013, p. 1, http:// d1c25a6gwz7q5e.cloudfront.net/reports/2013-10-01-Disasters-Leadership-Rebuilding .pdf.

229. Norimitsu Onishi and Martin Fackler, "Japanese Officials Ignored or Concealed Dangers," *New York Times*, May 16, 2011, http://www.nytimes.com/2011/05/17/ world/asia/17japan.html.

230. Initiative for Global Environmental Leadership, "Disasters, Leadership, and Rebuilding," 13.

231. "Fukushima 2013."

232. Kaufmann and Penciakova, "Japan's Triple Disaster."

233. Dick K. Nanto, William H. Cooper, J. Michael Donnelly, and Renée Johnson, "Japan's 2011 Earthquake and Tsunami: Economic Effects and Implications for the United States," Congressional Research Service Report for Congress, April 6, 2011, p. 1.

234. Suzuki and Kaneko, *Japan's Disaster Governance*, 7, 9.

235. Nanto, Cooper, Donnelly, and Johnson, "Japan's 2011 Earthquake and Tsunami," 1.

236. "Fukushima 2013."

237. Gill, Steger, and Slater, "3.11 Disasters," 21.

238. Ibid., 12.

239. Mike Adams, "Fukushima Victim Exposes Japanese Government's Attempted Bizarre Brainwashing of Radiation Victims," *Natural News*, August 21, 2013, http://www .naturalnews.com/041720_Fukushima_radiation_Japanese_government_propaganda _brainwashing.html.

240. Ibid.

241. National Diet of Japan, *The Official Report of the Fukushima Nuclear Accident Independent Investigation Commission*, 19–20.

242. Kaufmann and Penciakova, "Japan's Triple Disaster."

243. Junko Yoshida, "Failed Risk Analysis That Felled Fukushima," *EE Times*, March 8, 2016, http://www.eetimes.com/document.asp?doc_id=1329126.

244. "Fukushima 2013."

245. Suzuki and Kaneko, *Japan's Disaster Governance*, 64.

246. National Diet of Japan, *Official Report of the Fukushima Nuclear Accident Independent Investigation Commission*, 18–19.

247. Kaufmann and Penciakova, "Japan's Triple Disaster."

248. Ellis Krauss, "Crisis Management, LDP, and DPJ Style," *Japanese Journal of Political Science* 14 (2013): 191. See also Jeff Kingston, "Mismanaging Risk and the Fukushima Nuclear Crisis," *Japan Focus* 10 (2012), https://apjjf.org/2012/10/12/Jeff-Kingston/3724/article.html.

249. Initiative for Global Environmental Leadership, "Disasters, Leadership, and Rebuilding," 3.

250. Kaufmann and Penciakova, "Japan's Triple Disaster."

251. Initiative for Global Environmental Leadership, "Disasters, Leadership, and Rebuilding," 4.

252. Ibid.

253. Ibid., 5.

254. William Boardman, "Fukushima, a Global Conspiracy of Denial," *Reader Supported News*, January 3, 2014, http://readersupportednews.org/opinion2/271-38/21308-fukushima-a-global-conspiracy-of-denial.

255. Initiative for Global Environmental Leadership, "Disasters, Leadership, and Rebuilding," 5.

256. Geoff Brumfiel and Ichiko Fuyuno, "Fukushima's Legacy of Fear," *Nature* 483 (2012): 139.

257. DeMond Shondell Miller, "Public Trust in the Aftermath of Natural and Na-Technological Disasters," *International Journal of Sociology and Social Policy* 36 (2016): 420.

258. Initiative for Global Environmental Leadership, "Disasters, Leadership, and Rebuilding," 3.

259. Richard Hindmarsh, "Nuclear Disaster at Fukushima Daiichi: Introducing the Terrain," in *Nuclear Disaster at Fukushima Daiichi: Social, Political and Environmental Issues*, ed. Richard A. Hindmarsh (New York: Routledge, 2013), 6.

260. Richard Hindmarsh, "Fallout from Fukushima Daiichi: An Endnote," in *Nuclear Disaster at Fukushima Daiichi: Social, Political and Environmental Issues*, ed. Richard A. Hindmarsh (New York: Routledge, 2013), 215.

261. Initiative for Global Environmental Leadership, "Disasters, Leadership, and Rebuilding," 3.

262. Kaufmann and Penciakova, "Japan's Triple Disaster."

263. Boardman, "Fukushima, a Global Conspiracy of Denial."

264. Ibid.

265. Suzuki and Kaneko, *Japan's Disaster Governance*, 64.

266. Boardman, "Fukushima, a Global Conspiracy of Denial."

267. Linda Pentz Gunter, "No Bliss in This Ignorance: The Great Fukushima Nuclear Cover-Up," *The Ecologist*, February 20, 2016, http://www.theecologist.org/News/news_analysis/2987222/no_bliss_in_this_ignorance_the_great_fukushima_nuclear_coverup.html.

268. Robert Hunziker, "Fukushima Cover Up," *CounterPunch*, October 31, 2016, https://www.counterpunch.org/2016/10/31/fukushima-cover-up/.

269. Krauss, "Crisis Management, LDP, and DPJ Style," 192.

270. Gill, Steger, and Slater, "3.11 Disasters," 13.

271. Jeff Kingston, "Introduction," in *Natural Disaster and Nuclear Crisis in Japan: Response and Recovery after Japan's 3/11*, ed. Jeff Kingston (New York: Routledge, 2012), 7.

272. Katherine Harmon, "Japan's Post-Fukushima Earthquake Health Woes Go Beyond Radiation Effects," *Scientific American*, March 2, 2012, p. 1.

273. Gill, Steger, and Slater, "3.11 Disasters," 21.

274. Hindmarsh, "Nuclear Disaster at Fukushima Daiichi: Introducing the Terrain," 9.

275. Edward Group, "8 Shocking Health Effects from the Fukushima Disaster," Global Healing Center, June 9, 2014, http://www.globalhealingcenter.com/natural-health/health-effects-fukushima-disaster/.

276. Samuels, *3.11: Disaster and Change in Japan*, xi.

277. Gill, Steger, and Slater, "3.11 Disasters," 13.

278. Jordan Sand, "Living with Uncertainty After March 11, 2011," *Journal of Asian Studies* 71 (2012): 315.

279. National Diet of Japan, *Official Report of the Fukushima Nuclear Accident Independent Investigation Commission*, 19.

280. Gunter, "No Bliss in This Ignorance."

281. Greenpeace International, "Fukushima Citizens Remain Highly Exposed to Radiations," June 9, 2011, http://coastalcare.org/2011/06/fukushima-citizens-remain-highly-exposed-to-radiations/.

282. "Fukushima 2013."

283. Group, "8 Shocking Health Effects from the Fukushima Disaster."

284. Suzuki and Kaneko, *Japan's Disaster Governance*, 49.

285. Yoshida, "Failed Risk Analysis That Felled Fukushima."

286. Initiative for Global Environmental Leadership, "Disasters, Leadership, and Rebuilding," 7.

287. Ibid.

288. Sand, "Living with Uncertainty After March 11, 2011," 316.

289. Initiative for Global Environmental Leadership, "Disasters, Leadership, and Rebuilding," 8.

290. Hunziker, "Fukushima Cover Up."

291. Sasakawa Peace Foundation, "The Fukushima Nuclear Accident and Crisis Management: Lessons for Japan-U.S. Alliance Cooperation," September 2012, p. 7, https://www.spf.org/jpus/img/investigation/book_fukushima.pdf.

292. Jim Nichol, "Russia-Georgia Conflict in August 2008: Context and Implications for U.S. Interests," Congressional Research Service Report for Congress, March 3, 2009, p. 3.

293. Svante E. Cornell, Johanna Popjanevski, and Niklas Nilsson, *Russia's War in Georgia: Causes and Implications for Georgia and the World* (Washington, DC: Central Asia–Caucasus Institute and Silk Road Studies Program Joint Center, 2008), 5–13.

294. Charles King, "The Five-Day War: Managing Moscow After the Georgia Crisis," *Foreign Affairs* 87 (November–December 2009): 2.

295. "Five Years On, Georgia Makes Up with Russia," *BBC News*, June 25, 2013, http://www.bbc.co.uk/news/world-europe-23010526.

296. Nichol, "Russia-Georgia Conflict in August 2008," 15.

297. Brian J. Ellison, "Russian Grand Strategy in the South Ossetia War," *Demokratizatsiya* 19 (Fall 2011): 348.

298. Ellison, "Russian Grand Strategy in the South Ossetia War," 348–349.

299. Karl E. Meyer, "After Georgia: Back to the Future," *World Policy Journal* 25 (2008): 119.

300. Roy Allison, "Russia Resurgent? Moscow's Campaign to 'Coerce Georgia to Peace,'" *International Affairs* 84 (2008): 1145.

301. Meyer, "After Georgia: Back to the Future," 120, 122.

302. Ibid., 121.

303. King, "Five-Day War," 8.

304. Diana Digol, "Right or Wrong: Debate in Russia on Conflict in Georgia," *Security and Peace* 27 (2009): 112.

305. Allison, "Russia Resurgent?," 1145–1146.

306. Charles King, "Clarity in the Caucasus? The Facts and Future of the 2008 Russian-Georgian War," *Foreign Affairs*, October 11, 2009, https://www.foreignaffairs.com/articles/russia-fsu/2009-10-11/clarity-caucasus.

307. Meyer, "After Georgia," 119.

308. King, "Clarity in the Caucasus?"

309. Ibid.

310. David Pryce-Jones, "Cold War II," *National Review*, September 15, 2008, p. 24.

311. Ibid.

312. Cornell, Popjanevski, and Nilsson, *Russia's War in Georgia*, 25, 26.

313. King, "Five-Day War," 8–9.

314. Ibid.

315. Ariel Cohen and Robert E. Hamilton, *The Russian Military and the Georgia War: Lessons and Implications* (Carlisle, PA: Strategic Studies Institute, 2011), 2–4.

316. Nichol, *Russia-Georgia Conflict in August 2008*, 24.

317. Meyer, "After Georgia," 119.

318. Ibid., 121.

319. Ellison, "Russian Grand Strategy in the South Ossetia War," 346.

320. Ibid.

321. Meyer, "After Georgia," 119.

322. Cohen and Hamilton, *Russian Military and the Georgia War*, 6–7.

323. Allison, "Russia Resurgent?," 1149.

324. Ibid., 1163.

325. King, "Five-Day War," 6.

326. Pryce-Jones, "Cold War II," 24.

327. Robert P. Chatham, "Defense of Nationals Abroad: The Legitimacy of Russia's Invasion of Georgia," *Florida Journal of International Law* 23 (April 2011): 102.

328. King, "Five-Day War," 5–6.

329. Pryce-Jones, "Cold War II," 26.

330. Meyer, "After Georgia: Back to the Future," 120.

331. Pryce-Jones, "Cold War II," 26.

332. King, "Clarity in the Caucasus?"

333. King, "Five-Day War," 9.

334. Ibid., 10.

335. Allison, "Russia Resurgent?," 1161.

336. Ibid.

337. Cornell, Popjanevski, and Nilsson, *Russia's War in Georgia*, 31, 32.

338. Pryce-Jones, "Cold War II," 26.

339. King, "Five-Day War," 3–4.

340. Ibid., 7–8.

341. Ariel Cohen, James Carafano, and Lajos F. Szaszdi, "Russian Forces in the Georgian War: Preliminary Assessment and Recommendations," Heritage Foundation, August 20, 2008, https://www.heritage.org/europe/report/russian-forces-the-georgian-war-preliminary-assessment-and-recommendations.

342. Allison, "Russia Resurgent?," 1154.

343. Pryce-Jones, "Cold War II," 26.

344. King, "Five-Day War," 7–8.

345. Cohen and Hamilton, *Russian Military and the Georgia War*, 1.

346. Ibid., vii.

347. King, "Five-Day War," 8.

348. Ibid., 11.

349. Elliott Abrams, "Bombing the Syrian Reactor: The Untold Story," *Commentary*, February 1, 2013, http://www.commentarymagazine.com/article/bombing-the-syrian-reactor-the-untold-story/.

350. Daveed Gartenstein-Ross and Joshua D. Goodman, "The Attack on Syria's al-Kibar Nuclear Facility," *InFocus*, Spring 2009, http://www.jewishpolicycenter.org/826/the-attack-on-syrias-al-kibar-nuclear-facility.

351. Leonard S. Spector and Avner Cohen, "Israel's Airstrike on Syria's Reactor: Implications for the Nonproliferation Regime," *Arms Control Today* 38 (July–August 2008): 15.

352. Spector and Cohen, "Israel's Airstrike on Syria's Reactor," 15.

353. Abrams, "Bombing the Syrian Reactor."

354. Seymour M. Hersh, "A Strike in the Dark: What Did Israel Bomb in Syria?" *New Yorker*, February 11, 2008, http://www.newyorker.com/magazine/2008/02/11/a -strike-in-the-dark.

355. "Syria Target Hit by Israel Was 'Nuclear Site,'" *Al Jazeera*, April 29, 2011, http:// www.aljazeera.com/news/middleeast/2011/04/201142962917518797.html.

356. Anthony H. Cordesman, "Syrian Weapons of Mass Destruction: An Overview," Center for Strategic and International Studies, June 2, 2008, p. 3, https://www.tagesschau .de/ausland/syrien1850.pdf.

357. Ibid.,4–5.

358. Abrams, "Bombing the Syrian Reactor."

359. Ibid.

360. Ibid.

361. Hersh, "Strike in the Dark."

362. Ibid.

363. Richard Weitz, *Israeli Airstrike in Syria: International Reactions* (Monterey, CA: Monterey Institute for International Studies, 2007), 1.

364. Ibid., 1–2.

365. David E. Sanger and Mark Mazzetti, "Analysts Find Israel Struck a Syrian Nu-clear Project," *New York Times*, October 14, 2007, p. A1.

366. Gartenstein-Ross and Goodman, "Attack on Syria's al-Kibar Nuclear Facility."

367. Ibid.

368. Ibid.

369. Hersh, "Strike in the Dark."

370. Abrams, "Bombing the Syrian Reactor."

371. Ibid.

372. Ibid.

373. Elliott Abrams, Eliot Cohen, Eric Edelman, and John Hannah, "The Right Call on the Syrian Threat," *Washington Post*, September 15, 2011, http://www.washingtonpost .com/opinions/the-right-call-on-the-syrian-threat/2011/09/14/gIQAa85eVK_story.html.

374. Hersh, "Strike in the Dark."

375. Erich Follath and Holger Stark, "The Story of 'Operation Orchard': How Israel Destroyed Syria's Al Kibar Nuclear Reactor," *Spiegel Online*, November 2, 2009, http:// www.spiegel.de/international/world/0,1518,658663,00.html.

376. Abrams, "Bombing the Syrian Reactor."

377. Follath and Stark, "Story of 'Operation Orchard.'"

378. Hersh, "Strike in the Dark."

379. Riedel, "Lessons of the Syrian Reactor," 41.

380. Spector and Cohen, "Israel's Airstrike on Syria's Reactor," 15.

381. Andrew Garwood-Gowers, "Israel's Airstrike on Syria's Al-Kibar Facility: A Test Case for the Doctrine of Pre-emptive Self-Defence?" *Journal of Conflict and Security Law* 16 (2011): 263.

382. Spector and Cohen, "Israel's Airstrike on Syria's Reactor," 17.

383. Garwood-Gowers, "Israel's Airstrike on Syria's Al-Kibar Facility," 263–264.

384. Spector and Cohen, "Israel's Airstrike on Syria's Reactor," 17.

385. Hersh, "Strike in the Dark."

386. Follath and Stark, "Story of 'Operation Orchard.'"

387. Ibid.

388. David Makovsky, "The Silent Strike: How Israel Bombed a Syrian Nuclear Installation and Kept It Secret," *New Yorker*, September 17, 2012, http://www.newyorker .com/magazine/2012/09/17/the-silent-strike.

389. Gartenstein-Ross and Goodman, "Attack on Syria's al-Kibar Nuclear Facility."

390. Follath and Stark, "Story of 'Operation Orchard.'"

391. Hersh, "Strike in the Dark."

392. Peter Crail, "IAEA: Syrian Reactor Explanation Suspect," *Arms Control Today* 39 (March 2009): 45.

393. Garwood-Gowers, "Israel's Airstrike on Syria's Al-Kibar Facility," 270.

394. "Syria 'Had Covert Nuclear Scheme,'" *BBC News*, April 25, 2008, http://news .bbc.co.uk/2/hi/7364269.stm.

395. Riedel, "Lessons of the Syrian Reactor," 39.

396. Abrams, "Bombing the Syrian Reactor."

397. "George W Bush 'Considered Military Strike on Syria,'" *The Telegraph*, November 5, 2010, https://www.telegraph.co.uk/news/worldnews/us-politics/8113583/George -W-Bush-considered-military-strike-on-Syria.html.

398. Hersh, "Strike in the Dark."

399. "Syria 'Had Covert Nuclear Scheme.'"

400. Larry Derfner, "Israel Takes a Swipe at Syria," *U.S. News and World Report*, (October 1, 2007, p. 36.

401. Follath and Stark, "Story of 'Operation Orchard.'"

402. Ibid.

403. Gartenstein-Ross and Goodman, "Attack on Syria's al-Kibar Nuclear Facility."

404. Follath and Stark, "Story of 'Operation Orchard.'"

405. Hersh, "Strike in the Dark."

406. Abrams, "Bombing the Syrian Reactor."

407. "Are the Usual Suspects Responsible for Uzbekistan's Violence?" *EurasiaNet*, May 16, 2005, https://eurasianet.org/are-the-usual-suspects-responsible-for-uzbekistans -violence.

408. Sarah Kendzior, "Uzbekistan's Forgotten Massacre," *New York Times*, May 12, 2015, https://www.nytimes.com/2015/05/13/opinion/uzbekistans-forgotten-massacre.html.

409. "Are the Usual Suspects Responsible for Uzbekistan's Violence?"

410. Ibid.

411. Kendzior, "Uzbekistan's Forgotten Massacre."

412. Dilya Usmanova, "Andijan: A Policeman's Account," Institute for War and Peace, November 20, 2005, https://iwpr.net/global-voices/andijan-policemans-account-0.

413. "Waiting for the Other Shoe to Drop," *The Economist*, November 3, 2005, http://www.economist.com/node/5118919.

414. "Are the Usual Suspects Responsible for Uzbekistan's Violence?"

415. "Burying the Truth: Uzbekistan Rewrites the Story of the Andijan Massacre," *Human Rights Watch* 17 (September 2005): 2.

416. Stephen Blank, "Strategic Surprise? Central Asia in 2006," *China and Eurasia Forum Quarterly* 4 (May 2006): 109–110.

417. Ibid., 110.

418. Ibid.

419. Alex Rodriguez, "US Closes Air Base in Uzbekistan amid Uprising Dispute," *Boston Globe*, November 22, 2005, http://archive.boston.com/news/world/europe/articles/2005/11/22/us_closes_air_base_in_uzbekistan_amid_uprising_dispute/.

420. "Burying the Truth," 1–2.

421. "Waiting for the Other Shoe to Drop."

422. Kendzior, "Uzbekistan's Forgotten Massacre."

423. "Are the Usual Suspects Responsible for Uzbekistan's Violence?"

424. "Waiting for the Other Shoe to Drop."

425. "Are the Usual Suspects Responsible for Uzbekistan's Violence?"

426. "Waiting for the Other Shoe to Drop."

427. Kendzior, "Uzbekistan's Forgotten Massacre."

428. Ibid.

429. Blank, "Strategic Surprise?," 117.

430. Ibid., 110.

431. Ibid., 117.

432. "Are the Usual Suspects Responsible for Uzbekistan's Violence?"

433. Blank, "Strategic Surprise?," 110.

434. "Are the Usual Suspects Responsible for Uzbekistan's Violence?"

435. Kendzior, "Uzbekistan's Forgotten Massacre."

436. Blank, "Strategic Surprise?," 115.

437. Stephen Blank, "Uzbekistan: A Strategic Challenge to American Policy," Strategic Studies Institute, October 2004, p. 1, https://www.opensocietyfoundations.org/sites/default/files/blank_uzbekistan.pdf.

438. "Waiting for the Other Shoe to Drop."

439. Kendzior, "Uzbekistan's Forgotten Massacre."

440. Blank, "Uzbekistan," 20.

441. "Waiting for the Other Shoe to Drop."

442. Ibid.

443. Ibid.

444. Blank, "Strategic Surprise?," 124.

445. R. Jeffrey Smith and Glenn Kessler, "U.S. Opposed Calls at NATO for Probe of Uzbek Killings," *Washington Post*, June 14, 2005, http://www.washingtonpost.com/wp-dyn/content/article/2005/06/13/AR2005061301550.html.

446. "Are the Usual Suspects Responsible for Uzbekistan's Violence?"

447. Blank, "Strategic Surprise?," 125.

448. Ibid., 110.

449. "Are the Usual Suspects Responsible for Uzbekistan's Violence?"

450. Blank, "Strategic Surprise?," 109.

451. Chris Patten, "Saving Central Asia from Uzbekistan," *New York Times*, March 21, 2006, https://www.nytimes.com/2006/03/21/opinion/saving-central-asia-from-uzbekistan.html.

452. Blank, "Strategic Surprise?," 123, 126.

453. Condoleezza Rice, "Interview with Kubat Otorbaev of Radio Liberty and Sultan Jumagulov of the BBC Kyrgyz Service," U.S. Department of State, October 11, 2005, https://2001-2009.state.gov/secretary/rm/2005/54664.htm.

454. Blank, "Strategic Surprise?," 125.

455. Rodriguez, "US Closes Air Base in Uzbekistan amid Uprising Dispute."

456. Blank, "Strategic Surprise?," 112.

457. *The 9/11 Commission Report: Final Report of the National Commission on Terrorist Attacks upon the United States* (Washington, DC: U.S. Government Printing Office, 2004), 172, http://govinfo.library.unt.edu/911/report/911Report.pdf.

458. Gaetano Joe Ilardi, "The 9/11 Attacks: A Study of Al Qaeda's Use of Intelligence and Counterintelligence," *Studies in Conflict and Terrorism* 32 (2009): 171.

459. *9/11 Commission Report*, 341.

460. Ryan Thornton, "Changing the Game: Assessing Al Qaeda's Terrorist Strategy," *Harvard International Review* 27 (Fall 2005): 37.

461. Steve Jones, "US Foreign Policy After 9/11: Obvious Changes, Subtle Similarities," *ThoughtCo.*, April 17, 2018, http://usforeignpolicy.about.com/od/defense/a/Us-Foreign-Policy-After-9-11.htm.

462. Gallya Lahav, "Mobility and Border Security: The U.S. Aviation System, the State, and the Rise of Public-Private Partnerships," in *Politics at the Airport*, ed. Mark Salter (Minneapolis: University of Minnesota Press, 2008), 79.

463. Assaf Moghadam, "How Al Qaeda Innovates," *Security Studies* 22 (2013): 466–467.

464. Charles F. Parker and Eric K. Stern, "Blindsided? September 11 and the Origins of Strategic Surprise," *Political Psychology* 23 (September 2002): 601–602.

465. *9/11 Commission Report*, 341.

466. Ilardi, "9/11 Attacks," 177.

467. James Risen, "U.S. Failed to Act on Warnings in '98 of a Plane Attack," *New York Times*, September 19, 2002, p. 1.

468. Richard H. Shultz Jr. and Ruth Margolies Beitler, "Tactical Deception and Strategic Surprise in Al-Qai'da's Operations," *Middle East Review of International Affairs* 8 (June 2004): 57.

469. Ilardi, "9/11 Attacks," 179.

470. Shultz and Beitler, "Tactical Deception and Strategic Surprise in Al-Qai'da's Operations," 70, 71.

471. Ibid., 59.

472. Charles E. Parker and Eric K. Stern, "Bolt from the Blue or Avoidable Failure? Revisiting September 11 and the Origins of Strategic Surprise," *Foreign Policy Analysis* 1 (November 2005): 614.

473. Shultz and Beitler, "Tactical Deception and Strategic Surprise in Al-Qai'da's Operations," 58.

474. *9/11 Commission Report*, 345.

475. Daniel Byman, "Strategic Surprise and the September 11 Attacks," *Annual Review of Political Science* 8 (2005): 151.

476. Shultz and Beitler, "Tactical Deception and Strategic Surprise in Al-Qai'da's Operations," 56.

477. Jarret M. Brachman, "High-Tech Terror: Al-Qaeda's Use of New Technology," *Fletcher Forum of World Affairs* 30 (Summer 2006): 149, 151.

478. Ibid., 151, 153.

479. Ibid., 153.

480. Ibid., 156.

481. Parker and Stern, "Blindsided?," 606.

482. Ibid., 607.

483. *9/11 Commission Report*, 213.

484. Ibid., 340.

485. Gerald Posner, *Why America Slept: The Failure to Prevent 9/11* (New York: Random House, 2003), 3.

486. Parker and Stern, "Blindsided?," 607.

487. Posner, *Why America Slept*, 147.

488. *9/11 Commission Report*, 342–343.

489. Posner, *Why America Slept*, 153.

490. *9/11 Commission Report*, 341.

491. Ilardi, "9/11 Attacks," 182.

492. Byman, "Strategic Surprise and the September 11 Attacks," 165–166.

493. Dan Caldwell and Robert E. Williams Jr., *Seeking Security in an Insecure World* (Lanham, MD: Rowman and Littlefield, 2006), 170.

494. Byman, "Strategic Surprise and the September 11 Attacks," 145.

495. Ilardi, "9/11 Attacks," 181.

496. Ibid.

497. Daniel S. Harawa, "The Post-TSA Airport: A Constitution Free Zone?" *Pepperdine Law Review* 41 (2013): 11–12.

498. Richard N. Haass, "Terrorism Concerns After Bin Laden," *Huffington Post*, May 2, 2011, https://www.huffingtonpost.com/richard-n-haass/terrorism-concerns-after-_b_856253.html.

499. Lee Ferran and Rym Momtaz, "ISIS: Trail of Terror," *ABC News*, February 23, 2015, http://abcnews.go.com/WN/fullpage/isis-trail-terror-isis-threat-us-25053190.

500. "How Should Nations Respond to the ISIS Threat?" Wharton School, November 17, 2015, http://knowledge.wharton.upenn.edu/article/how-should-nations-respond-to-the-isis-threat/.

501. Brendan I. Koerner, "Why ISIS Is Winning the Social Media War," *Wired*, April 2016, https://www.wired.com/2016/03/isis-winning-social-media-war-heres-beat/.

502. "Attacks Draw Mixed Response in Mideast," *CNN*, September 12, 2001, http://www.cnn.com/2001/WORLD/europe/09/12/mideast.reaction.

503. Brachman, "High-Tech Terror, 152.

504. *9/11 Commission Report*, xvi.

505. Posner, *Why America Slept*, xi.

506. Dean Schabner and Karen Travers, "Osama bin Laden Killed: 'Justice Is Done,' President Says," *ABC News*, May 1, 2011, http://abcnews.go.com/Blotter/osama-bin-laden-killed/story?id=13505703.

507. "Crowds Gather at Ground Zero, White House to Cheer bin Laden's Death," *New York Post*, May 2, 2011, https://nypost.com/2011/05/02/crowds-gather-at-ground-zero-white-house-to-cheer-bin-ladens-death.

508. David Danielo, "Bin Laden's Death and the Moral Level of War," Foreign Policy Research Institute, May 3, 2011, https://www.fpri.org/article/2011/05/more-fpri-perspectives-on-bin-ladens-demise.

509. Reuters, "Bin Laden Was Found at Luxurious Pakistan Compound," *CNBC*, May 2, 2011, https://www.cnbc.com/id/42854159.

510. Ed Husain, "Bin Laden as 'Martyr': A Call to Jihadists," *CNN*, May 4, 2011, http://www.cnn.com/2011/OPINION/05/02/husain.bin.laden.

511. *9/11 Commission Report*, 170.

512. Parker and Stern, "Bolt from the Blue or Avoidable Failure?," 323.

513. "US 'Losing Media War to al-Qaeda,'" *BBC News*, February 17, 2006, http://news.bbc.co.uk/2/hi/americas/4725992.stm.

514. Brachman, "High-Tech Terror," 150.

515. Dave Johns, "The Crimes of Saddam Hussein: 1990; The Invasion of Kuwait," *PBS*, January 24, 2006, http://www.pbs.org/frontlineworld/stories/iraq501/events_kuwait.html.

516. Ibid.

517. Christopher Greenwood, "New World Order or Old? The Invasion of Kuwait and the Rule of Law," *Modern Law Review* 55 (March 1992): 154.

518. Joe Stork and Ann M. Lesch, "Background to the Crisis: Why War?" *Middle East Report*, no. 167 (1990): 11.

519. William J. Perry, "Desert Storm and Deterrence," *Foreign Affairs* 70 (Fall 1991): 66–82.

520. Eliot A. Cohen, "A Revolution in Warfare," *Foreign Affairs* 75 (March–April 1996): 30–40.

521. John Mueller, "The Perfect Enemy: Assessing the Gulf War," *Security Studies* 5 (Autumn 1995): 106; Jeffrey Record, *Hollow Victory* (Washington, DC: Brassey's, 1993), 6, 135.

522. Michael Wines, "The Iraqi Invasion: U.S. Says Bush Was Surprised by the Iraqi Strike," *New York Times*, August 5, 1990, http://www.nytimes.com/1990/08/05/world/the-iraqi-invasion-us-says-bush-was-surprised-by-the-iraqi-strike.html.

523. Musallam Ali Musallah, *The Iraqi Invasion of Kuwait* (London: British Academic Press, 1996), 4.

524. Association for Diplomatic Studies and Training, "A Day of Mixed Messages over Iraq's Invasion of Kuwait," May 4, 2017, http://adst.org/2017/05/day-mixed-messages-iraqs-invasion-kuwait/.

525. Duelfer and Dyson, "Chronic Misperception and International Conflict," 86.

526. Ibid., 85–86; Efraim Karsh and Inari Rautsi, *Saddam Hussein: A Political Biography* (New York: Free Press, 1991), 216; George H. W. Bush and Brent Scowcroft, *A World Transformed* (New York: Knopf, 1998), 311.

527. Strauss, "Attacking Iraq," 15.

528. Stork and Lesch, "Background to the Crisis," 18.

529. Charles E. Allen, "Warning and Iraq's Invasion of Kuwait: A Retrospective Look," *Defense Intelligence Journal* 7 (1998): 37–38.

530. Wines, "Iraqi Invasion."

531. Ibid.

532. Kevin M. Woods and Mark E. Stout, "Saddam's Perceptions and Misperceptions: The Case of Desert Storm," *Journal of Strategic Studies* 33 (2010): 7–9; Wines, "Iraqi Invasion."

533. Allen, "Warning and Iraq's Invasion of Kuwait," 37–38.

534. Wines, "Iraqi Invasion."

535. J. E. McReynolds, "Kuwait Helpless Against Iraq, Analyst Says," *NewsOK*, August 3, 1990, http://newsok.com/article/2326074.

536. "Iraq Invades Kuwait," *BBC News*, August 2, 1990, http://news.bbc.co.uk/onthisday/hi/dates/stories/august/2/newsid_2526000/2526937.stm.

537. Thomas C. Hayes, "Confrontation in the Gulf: The Oilfield Lying Below the Iraq-Kuwait Dispute," *New York Times*, September 3, 1990, http://www.nytimes.com/1990/09/03/world/confrontation-in-the-gulf-the-oilfield-lying-below-the-iraq-kuwait-dispute.html.

538. Stork and Lesch, "Background to the Crisis," 18.

539. Ibid., 12.

540. Hayes, "Confrontation in the Gulf."

541. Stork and Lesch, "Background to the Crisis," 11.

542. Hayes, "Confrontation in the Gulf."

543. Stork and Lesch, "Background to the Crisis," 18.

544. Johns, "Crimes of Saddam Hussein."

545. Stork and Lesch, "Background to the Crisis," 11, 15.

546. Ibid., 11.

547. See, for example, ibid.

548. Ibid.

549. Hamdi A. Hassan, *The Iraqi Invasion of Kuwait* (London: Pluto, 1999), 4.

550. Woods and Stout, "Saddam's Perceptions and Misperceptions," 6.

551. Johns, "Crimes of Saddam Hussein."

552. Greenwood, "New World Order or Old?," 178.

553. Ibid., 153.

554. Jerry M. Long, *Saddam's War of Words* (Austin: University of Texas Press, 2004), 24.

555. "Iraq Invades Kuwait."

556. Johns, "Crimes of Saddam Hussein."

557. Ibrahim Al-Marashi, "Saddam's Security Apparatus During the Invasion of Kuwait and the Kuwaiti Resistance," *Journal of Intelligence History* 3 (Winter 2003): 61.

558. Duelfer and Dyson, "Chronic Misperception and International Conflict," 92–93.

559. Strauss, "Attacking Iraq," 15.

560. Ibid.

561. Kenneth John, "Lessons of the Persian Gulf War," *Freedom Rings Radio*, http://www.freedomrings.net/html/writings/essays/Lessons_of_the_Persian_Gulf_War.htm.

562. Stork and Lesch, "Background to the Crisis," 18.

Chapter 4

1. Jon Latimer, *Deception in War* (New York: Overlook, 2001), 129.

2. Yaacov Y. I. Vertzberger, *The World in Their Minds: Information Processing, Cognition, and Perception in Foreign Policy Decisionmaking* (Stanford, CA: Stanford University Press, 1990), 17.

3. Lani Kass and J. Phillip London, "Surprise, Deception, Denial and Warning: Strategic Imperatives," *Orbis* 57 (Winter 2013): 67.

Chapter 5

1. Ephraim Kam, *Surprise Attack: The Victim's Perspective* (Cambridge, MA: Harvard University Press, 1988), 53.

2. Richards J. Heuer, *Psychology of Intelligence Analysis* (Washington, DC: Central Intelligence Agency Center for the Study of Intelligence, 1999), 6.

3. Cynthia M. Grabo, *Anticipating Surprise: Analysis for Strategic Warning* (Bethesda, MD: Center for Strategic Intelligence Research at the Joint Military Intelligence College, 2002), 133.

4. James Gleick, *The Information: A History, a Theory, a Flood* (New York: Pantheon, 2011), 403.

5. "Too Much Information: How to Cope with Data Overload," *The Economist*, June 30, 2011, http://www.economist.com/node/18895468

6. Gleick, *Information*, 409, 410.

7. "Too Much Information."

8. Ibid.

9. Ibid.

10. Heuer, *Psychology of Intelligence Analysis*, 175.

11. Michael Webb, *Creating a New Reality: Information Age Effects on the Deception Process* (Maxwell Air Force Base, Alabama: Air University School of Advanced Air and Space Studies, 2006), 1.

12. Jeffrey Record, "Back to the Weinberger-Powell Doctrine?" *Strategic Studies Quarterly*, Fall 2007, p. 91.

13. Richard H. Shultz Jr. and Ruth Margolies Beitler, "Tactical Deception and Strategic Surprise in Al-Qaïda's Operations," *Middle East Review of International Affairs* 8 (June 2004): 61.

14. M. R. D. Foot, "Conditions Making for Success and Failure of Denial and Deception: Democratic Regimes," in *Strategic Denial and Deception: The Twenty-First Century Challenge*, ed. Roy Godson and James J. Wirtz (New Brunswick, NJ: Transaction, 2002) 122.

15. Magnus Christiansson, *Strategic Surprise in the Ukraine Crisis* (Stockholm, Sweden: Swedish National Defence College, 2014), 61.

16. Jakub Grygiel, "The Return of Europe's Nation-States: The Upside to the EU's Crisis," *Foreign Affairs* 95 (2016): 95.

17. Lani Kass and J. Phillip London, "Surprise, Deception, Denial and Warning: Strategic Imperatives," *Orbis* 57 (Winter 2013): 67.

18. Grabo, *Anticipating Surprise*, 45–46.

19. Michael I. Handel, "Intelligence and the Problem of Strategic Surprise," *Journal of Strategic Studies* 7 (1984): 239.

20. Grabo, *Anticipating Surprise*, 135.

21. Ibid., 21.

22. Kass and London, "Surprise, Deception, Denial and Warning," 76-77.

23. Yaacov Y. I. Vertzberger, *The World in Their Minds: Information Processing, Cognition, and Perception in Foreign Policy Decisionmaking* (Stanford, CA: Stanford University Press, 1990), 357.

24. Ibid., 360.

25. Grabo, *Anticipating Surprise*, 130.

26. Ibid., 131.

27. *The 9/11 Commission Report: Final Report of the National Commission on Terrorist Attacks upon the United States* (Washington, DC: U.S. Government Printing Office, 2004), 388.

28. Assaf Moghadam, "How Al Qaeda Innovates," *Security Studies* 22 (2013): 467.

29. Ariel Cohen and Robert E. Hamilton, *The Russian Military and the Georgia War: Lessons and Implications* (Carlisle, PA: Strategic Studies Institute, 2011), 15.

30. Grabo, *Anticipating Surprise*, 129–130.

31. Roy Godson and James J. Wirtz, "Strategic Denial and Deception," *International Journal of Intelligence and CounterIntelligence* 13 (2000): 425–426.

32. Timothy J. Smith, "Overlord/Bodyguard: Intelligence Failure Through Adversary Deception," *International Journal of Intelligence and CounterIntelligence* 27 (2014): 555.

33. Godson and Wirtz, "Strategic Denial and Deception," 425–426.

34. Shultz and Beitler, "Tactical Deception and Strategic Surprise," 61.

35. Bruce Schneier, *Beyond Fear: Thinking Sensibly About Security in an Uncertain World* (New York: Copernicus, 2003), 215.

36. Smith, "Overlord/Bodyguard," 555.

37. Shultz and Beitler, "Tactical Deception and Strategic Surprise," 62.

38. Robert Mandel, *Optimizing Cyberdeterrence: A Comprehensive Strategy for Preventing Foreign Cyberattacks* (Washington, DC: Georgetown University Press, 2017), 203.

39. Gerald Sussman, *Communication, Technology, and Politics in the Information Age* (Thousand Oaks, CA: Sage, 1997), 264.

40. Godson and Wirtz, "Strategic Denial and Deception," 425–426.

41. Ibid.

42. Kass and London, "Surprise, Deception, Denial and Warning," 67.

43. Michael I. Handel, "Intelligence and Deception," in *Military Deception and Strategic Surprise*, ed. John Gooch and Amos Perlmutter (Totawa, NJ: Cass, 1982) 137.

44. Kass and London, "Surprise, Deception, Denial and Warning," 66.

45. Godson and Wirtz, "Strategic Denial and Deception," 431.

46. Brian Michael Jenkins, *Unconquerable Nation: Knowing Our Enemy, Strengthening Ourselves* (Santa Monica, CA: Rand, 2006), 53.

47. Godson and Wirtz, "Strategic Denial and Deception," 431.

48. Kass and London, "Surprise, Deception, Denial and Warning," 67.

49. Godson and Wirtz, "Strategic Denial and Deception," 431.

50. Grabo, *Anticipating Surprise*, 155.

51. Charles A. Duelfer and Stephen Benedict Dyson, "Chronic Misperception and International Conflict: The U.S.-Iraq Experience," *International Security* 36 (Summer 2011): 76.

52. Jamie Holmes, *Nonsense: The Power of Not Knowing* (New York: Crown, 2015), 10, 14.

53. Heuer, *Psychology of Intelligence Analysis*, 66.

54. Holmes, *Nonsense: The Power of Not Knowing*, 9–14.

55. Daniel J. Levitin, *The Organized Mind: Thinking Straight in the Age of Information Overload* (New York: Penguin, 2014), 383.

56. Joint Chiefs of Staff, "Military Deception," Joint Publication 3-13.4, July 13, 2006, p. A-1.

57. Holmes, *Nonsense: The Power of Not Knowing*, 15.

58. Shultz and Beitler, "Tactical Deception and Strategic Surprise," 61.

59. Gaetano Joe Ilardi, "The 9/11 Attacks: A Study of Al Qaeda's Use of Intelligence and Counterintelligence," *Studies in Conflict and Terrorism* 32 (2009): 179.

60. See Robert Mandel, *Coercing Compliance: State-Initiated Brute Force in Today's World* (Stanford, CA: Stanford University Press, 2015), 6–9.

61. Foot, "Conditions Making for Success and Failure," 117.

62. Daniel Byman, "Strategic Surprise and the September 11 Attacks," *Annual Review of Political Science* 8 (2005): 150.

63. Richard K. Betts, *Surprise Attack: Lessons for Defense Planning* (Washington, DC: Brookings Institution, 1982), 285, 295, 309–310.

64. Grabo, *Anticipating Surprise*, 129.

65. John Hollister Hedley, "The Challenges of Intelligence Analysis," in *Strategic Intelligence I: Understanding the Hidden Side of Government*, ed. Loch K. Johnson (Westport, CT: Praeger Security International, 2007), 124.

66. Heuer, *Psychology of Intelligence Analysis*, 74.

67. Charles F. Parker and Eric K. Stern, "Blindsided? September 11 and the Origins of Strategic Surprise," *Political Psychology* 23 (September 2002): 624.

68. Handel, "Intelligence and the Problem of Strategic Surprise," 245.

69. See Betts, *Surprise Attack*.

70. Thomas Quiggin, *Seeing the Invisible: National Security Intelligence in an Uncertain Age* (Singapore: World Scientific, 2007), 55.

71. *9/11 Commission Report*, 346.

72. Jack Davis, *Strategic Warning: If Surprise Is Inevitable, What Role for Analysis?* (Washington, DC: Central Intelligence Agency Sherman Kent Center, 2003), 6–16.

73. Kelvin Fu Wen Hao, "Globalisation and Its Impact on Military Intelligence," *Pointer: Journal of the Singapore Armed Forces* 41 (2015): 4.

74. *9/11 Commission Report*, 347–348.

75. Parker and Stern, "Blindsided?," 624.

76. *9/11 Commission Report*, 406.

77. Moghadam, "How Al Qaeda Innovates," 467.

78. Parker and Stern, "Blindsided?," 622.

79. Janne E. Nolan and Douglas MacEachin, *Discourse, Dissent, and Strategic Surprise: Formulating U.S. Security Policy in an Age of Uncertainty* (Washington, DC: Georgetown University Institute for the Study of Diplomacy, 2006), 109–110.

80. Heuer, *Psychology of Intelligence Analysis*, 76.

Conclusion

1. Lani Kass and J. Phillip London, "Surprise, Deception, Denial and Warning: Strategic Imperatives," *Orbis* 57 (Winter 2013): 81.

2. See Andrew M. Scott, *The Revolution in Statecraft: Informal Penetration* (New York: Random House, 1965).

3. Joint Chiefs of Staff, "Military Deception," Joint Publication 3-13.4, July 13, 2006, p. III-8.

4. Robert Jervis, *The Logic of Images in International Relations* (New York: Columbia University Press, 1989), 43, 63.

5. Yaacov Y. I. Vertzberger, *The World in Their Minds: Information Processing, Cognition, and Perception in Foreign Policy Decisionmaking* (Stanford, CA: Stanford University Press, 1990), 27.

6. Bernard I. Finel and Kristin M. Lord, "The Surprising Logic of Transparency," *International Studies Quarterly* 43 (June 1999): 316, 317, 320, 335.

7. Tyler Cowen, "How Real News Is Worse Than Fake News," *Bloomberg*, September 5, 2018, https://www.bloomberg.com/view/articles/2018-09-05/how-real-news-is-worse-than-fake-news; Finel and Lord, "The Surprising Logic of Transparency," 320, 335.

8. William J. Broad, "U.S. Rethinks Strategy for the Unthinkable," *New York Times*, December 15, 2010, http://www.nytimes.com/2010/12/16/science/16terror.html.

9. Quentin Hardy, "Why Big Data Is Not Truth," *New York Times*, June 1, 2013, http://bits.blogs.nytimes.com/2013/06/01/why-big-data-is-not-truth/.

10. Anthony Oettinger, "Information Overload: Managing Intelligence Technologies," *Harvard International Review*, September 6, 2002, http://hir.harvard.edu/article/?a=1061.

11. Anthony G. Oettinger, "Whence and Whither Intelligence, Command and Control? The Certainty of Uncertainty," Harvard University Program on Information Resources Policy, February 1990, p. 5, http://www.pirp.harvard.edu/pubs_pdf/oetting/oetting-p90-1.pdf.

12. Drew Conway, "Data Science in the U.S. Intelligence Community," *IQT Quarterly* 2 (Spring 2011): 27.

13. Vertzberger, *World in Their Minds*, 26–27. See also Kelvin Fu Wen Hao, "Globalisation and Its Impact on Military Intelligence," *Pointer: Journal of the Singapore Armed Forces* 41 (2015): 1; and Daniel S. Gressang IV, "The Shortest Distance Between Two Points Lies in Rethinking the Question: Intelligence and the Information Age Technology Challenge," in *Strategic Intelligence: Understanding the Hidden Side of Government*, ed. Loch K. Johnson (Westport, CT: Praeger Security International, 2007), 132.

14. Bill Kovach and Tom Rosenstiel, *Blur: How to Know What's True in the Age of Information Overload* (Bloomsbury, NJ: Bloomsbury USA, 2011), 7, 201.

Index

Threat: ambiguity of, 13, 32, 80; dangers associated with underreaction and over-reaction to, 28; difficulties assessing, 23, 28, 187; erosion of government com-munication during, 16; expanding range of, 15–16; as justification for existence of the state, 26; and need to project into the future, 167; overwhelming threat data, 20; posed by global data shock, 17, 45, 50, 51, 200; posed by information overload, 8; proliferation of transna-tional, 19

Transparency: appearance of, 55; citizens' desire for, from government, 19, 48, 51, 65, 104, 197; controversies over, 3, 98, 151, 180, 197, 198; and credibility, 13; dangers associated with, 52, 199; hypocrisy and, 199–200; and information collection, 17–18; information transparency paradox, 198–200; institutional, 6; as open to mis-interpretation, 163–164, 198; as promoting trust, 6, 161, 162, 191; as reinforced by spread of democracy, 6; safety of, 199; and strategic manipulation, 45, 52, 197; and value of an open press, 67

Trump, Donald, 59–67

Trust: blind trust by target of manipulation initiator, 160; enlightened internationalist expectations of, 1, 111; falling confidence among manipulation targets, 42, 87–88, 91, 101–102, 103, 104, 111, 133, 151, 156, 181, 197; in global security data, 22–23; and information explosion, 6; in informa-tion received, 15, 45, 202; loss of, among allies, 45, 47, 51, 52–53, 66, 73, 83, 91, 103; as promoted by transparency, 6, 198; as reduced for governments by mass public information access, 9; in value of big data analysis, 13, 200; in value of strategic manipulation, 63

United States: attractiveness of, as a manipu-lation target, 50, 128, 131–132, 192, 193; aversion of, to manipulation use, 33; and information misinterpretation, 77, 79–80, 84, 103; loss of global credibility of, 62, 64, 67, 74, 81, 82, 83, 108, 110, 111, 112, 127, 133–134, 136; need of, to improve manipu-lation defenses, 50, 52, 82, 137, 180, 183, 192; tension of, with allies, 62, 66, 82, 92, 103, 111, 112, 126–127, 136, 143–144, 153–154;

as victim of surprise, 83, 84, 113, 121, 127, 129–130, 131, 132, 138–139, 140

Unpredictability: benefits from, 35, 187–188; as consequence of strategic ambiguity, deception, and surprise, 35, 45, 47; con-troversies over, 3, 24, 66, 67; as embedded in Trump's foreign policy style, 59–62, 65–66, 171

Vigilance: in early-warning systems, 165; hypervigilance by targets, 53; in improv-ing information interpretation, 185; in managing global data shock, 167; and massive defense spending, 137; as needed to defend low-probability high-impact threats, 188; as needed to defend strategic manipulation, 22, 23, 41, 158, 160, 176, 181, 196; in postmanipulation recovery, 185; strategic manipulation against vigilant targets, 45

Violence: and absence of mutual trust and respect, 197; and action-reaction cycles, 53, 119, 151; illegitimacy of, 125–126; and intolerance, 162; low payoffs of, 81–82, 142; manipulating images of, 122; persis-tence of, over time, 75, 129, 144; and plac-ing blame on radical Islamic extremists, 122–123, 136; probability of deception fail-ing against violent groups, 52; regional contagion of, 126, 151, 154; use of social media to promote, 134–135

Vulnerability to manipulation and disruption: of citizens and state, 6, 8, 9, 16, 133–134, 135, 156, 158, 160, 175–176, 180, 194, 201, 202; and dangers of transparency, 199; of digital security system, 8–9, 20, 21, 45, 131; under information overload, 151–152; most vulnerable targets, 31, 159–160, 164; and need to project into the future, 167; reduction of one as increasing others, 172

War: ambiguity among war initiators, 46; prevalence of surprise military attack commencing, 44; unreliability of data during, 22–23

Wariness, 22, 23, 111, 150, 166, 185, 186

Warning, strategic: dangers of underreac-tion and overreaction to, 28, 181; deadly nature of, 38; and difficulties distinguish-ing signals from noise, 44, 130, 136; and